BEST LITTLE BOOK OF BIRDS

Frontispiece: Herring Gull

Photo and illustration credits appear on page 354.

Timber Press
Workman Publishing
Hachette Book Group, Inc.
1290 Avenue of the Americas
New York, New York 10104
timberpress.com

Timber Press is an imprint of Workman Publishing, a division of Hachette Book Group, Inc. The Timber Press name and logo are registered trademarks of Hachette Book Group, Inc.

Second printing 2026
Printed in Dongguan, China, (TLF) on responsibly sourced paper

Text design by Joel Ruffier based on series design by Vincent James
Jacket design by Vincent James

ISBN 978-1-64326-381-6

A catalog record for this book is available from the Library of Congress.

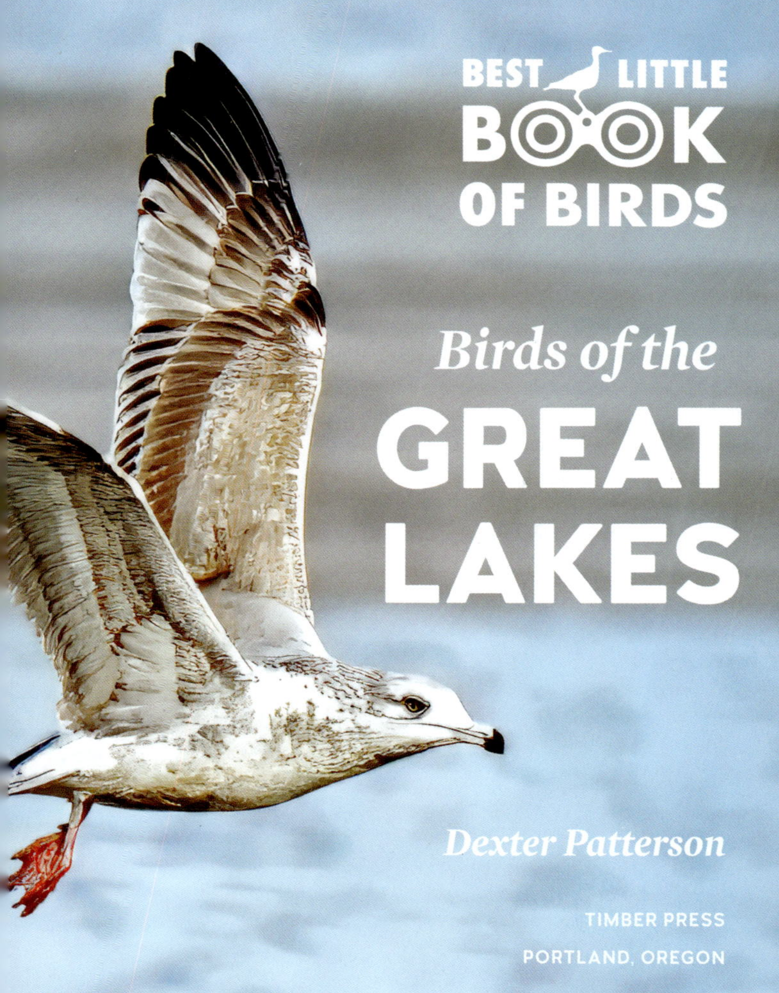

BEST LITTLE BOOK OF BIRDS

Birds of the GREAT LAKES

Dexter Patterson

TIMBER PRESS
PORTLAND, OREGON

CONTENTS

Preface

Welcome to the enchanting world of birdwatching in the Great Lakes region. My name is Dexter Patterson, and I am an educator, a multimedia professional, and cofounder of the BIPOC (Black, Indigenous, People of Color) Birding Club of Wisconsin. Our club is a vibrant community of people who share a love for the outdoors, the natural beauty of Wisconsin, and the diverse birdlife that graces our state. We welcome BIPOC birders and allies of all skill levels to join us on our field trips and events, fostering connections and shared experiences for our members.

I believe that birding is for everyone, and this perspective has deepened my passion for birds—a legacy ignited by my late grandfather, Ed Brown. Grandpa Brown instilled a simple yet profound mantra in his family: pay attention, and the birds will reveal themselves to you. Observing birds is magical, whether you're an experienced birder or a curious beginner. I hope to share some of that magic with you in these pages.

This book is not just a guide but a companion on your journey into birdwatching. It's designed to make birdwatching accessible to all, providing valuable tips, essential tools, and insightful perspectives to help you expand your knowledge as a birder. While this compilation doesn't cover every bird that visits or resides in the Great Lakes region, it does provide a window into over 150 species spanning 40 families—more than enough to get you excited!

So let's spread our wings and dive into this adventure with open hearts and curious minds. Who's ready to discover the beauty of birding in the Great Lakes region? Let's go!

Birding in the Great Lakes Region

Birdwatching in the Great Lakes Region is more than just a hobby—it's a front-row seat to one of nature's greatest spectacles. Each season brings new discoveries, from the vibrant activity of spring migration along the waterways and shores of the Great Lakes to the quiet resilience of wintering owls in the Northwoods of Wisconsin. Everyone can enjoy birdwatching, but having a little knowledge about common birds—their appearances, behaviors, and preferred habitats—can transform a simple walk in a local park into an unforgettable birding experience.

Tools and Tips

Embarking on a journey into the birding world requires more than just enthusiasm; it requires the right tools to enhance your experience and knowledge.

Use binoculars: Good optics bridge the gap between you and birds, offering a closer look at their plumage, behavior, and habitats. Invest in binoculars that are comfortable to use and within your budget. When shopping, consider the 8×32 or 10×42 magnification for optimal clarity and detail. Most birding clubs provide binoculars you can borrow at events, and don't hesitate to ask other birders about their optics and what they like and don't like.

Use a field guide: Field guides are indispensable for delving deeper into bird identification. Beyond the regional guide you hold in your hands, you might want to invest in a guide to a wider portion of North America or even a guide to a specific group of birds, perhaps warblers, waterfowl, or gulls. *The Sibley Field Guide to Birds of Eastern North America* and the *Peterson Field Guide to Birds of Eastern and Central North America* are my favorites.

Learn bird calls and songs: You will hear many birds before you see them, so familiarizing yourself with bird sounds will make you more aware of their presence. When you see a singing bird, spend time watching and listening to identify the bird and link it with its voice. Learn one bird and one call at a time; you don't need to learn them all right away.

Use birding apps: Apps include illustrations, sounds, and sometimes video, and cover everything from warblers to raptors to all the species in North America. The free Merlin Bird ID app is indispensable for learning bird songs and identifying birds you've never encountered before, and the free Raptor ID app will tell you all you need to know about birds of prey. Incorporating apps into your birding toolkit enhances your observational skills and expands your bird knowledge base.

Use a camera or smartphone: Every birding encounter is a moment to cherish, and capturing these fleeting glimpses enriches your journey. Whether with a dedicated camera or the convenience of a smartphone, documenting your sightings immortalizes the magic of birding. Share these moments with loved ones and relive the thrill of each discovery through your imagery. Don't worry about the photo quality; focus and enjoy the moment.

Join a local birding group: Birding with a group is a great way to learn from experienced birders and connect with others who share your interests.

Use local resources: Birding Backpack programs offered through libraries or community centers lend birding equipment, ensuring that everyone can go birdwatching without financial constraints. Find out if you have a program like this in your area!

Find a good birding location: Your backyard, neighborhood, or porch can be a great location to spot various birds. You can also explore places near you that are known for birdwatching, such as local parks, wetlands, and wildlife refuges.

Keep a birding journal: Tracking your bird sightings and observations can help you remember what you've seen and improve your identification skills.

Be patient: Birds can be elusive, so be prepared to spend time searching for them and enjoy the process of looking and listening. Sometimes, you won't see what you're looking for, but you will always see *something*.

Respect wildlife: Always observe birds from a distance and avoid disturbing them. Never chase or harass birds; always be nice to our feathered friends.

Enjoy yourself: Birding should be fun! It doesn't matter if you're just starting out; birding can offer a fun and meaningful way to explore the natural world.

With these essential tools and tips, your birding quest transforms into a voyage of discovery and connection with birds. As you explore diverse habitats and observe feathered marvels, let these tools be your companions in unraveling some of the complexity of birdwatching.

Using This Book

This book is your practical companion, offering insights and tools to enhance your birding experiences. It focuses on over 150 bird species found in the Great Lakes region, whether year-round residents, seasonal visitors, or passing migrants during fall or spring. The classification system used by scientists worldwide categorizes living things into smaller and smaller related groups, the last three being family, genus, and species. In this section you will find an introduction to the families featured in the book. Individual bird entries will include its genus and species. The language of classification is Latin, and an organism's genus and species are sometimes referred to as its "Latin name." In a departure from the typical field guide format that groups birds solely by genetic relationships, this book groups bird families that are genetically unrelated but resemble each other and share the same habitat. For example, the unrelated geese, swans, ducks, grebes, and loons that you might see swimming on a lake have been placed on adjacent pages for easier comparison.

Each species entry provides the following specifics:

PHOTOS
to help you to confirm your identification visually and appreciate the beauty of each species

CAPTION
mentions the sex and characteristics of the species shown

COMMON NAME

Latin Name

APPEARANCE Size, plumage color, bill shape, and other diagnostic markings. These descriptions are essential for precise identification in the field.

HABITAT Information about preferred habitat, such as forests, wetlands, gardens, or water, which gives you a better understanding of where to look for them.

BEHAVIOR Observations on typical feeding habits, nesting practices, flight patterns, or social interactions. This knowledge will enhance your ability to locate and observe them in their natural habitats.

FUN FACT Each profile includes an interesting tidbit about the bird, adding an element of fun to your learning. These facts may highlight unique behaviors, historical anecdotes, or quirky traits that make each species unique.

LENGTH HEAD TO TAIL / WINGSPAN / FAMILY SILHOUETTE

Beyond the Book

Birdwatching is a lifelong journey filled with moments of awe and wonder. While this book offers a window into the diversity of birds in the Great Lakes region, it is a partial list of all the species you might encounter. Use it as a starting point to develop your birding skills and as an invitation to explore further. Keep a birding journal, participate in local bird counts or citizen science projects, and join birding communities to continue your education.

A Year in the Life of a Birdwatcher in the Great Lakes Region

Birdwatching in the Great Lakes region is a fluid, year-round adventure shaped by the ebb and flow of four unique seasons. Each brings its own rhythm, species, and behaviors, offering birdwatchers endless opportunities to observe and connect with amazing birds.

WINTER

Winter can be a time of quiet solitude, for both birds and birdwatchers. It's cold, and the landscape is often covered in snow. Many species have migrated south, leaving behind a hardy group of winter residents and a few new arrivals. However, the remaining birds are often easier to spot against the leafless trees and snow-covered ground.

Some key species to look for during winter include the Northern Cardinal, Black-capped Chickadee, Dark-eyed Junco, and American Tree Sparrow. These birds are adapted to the cold, and their behaviors—such as flocking and frequent feeder visits—offer a close-up view. One of my favorite sights is spotting a bright red cardinal against a snowy backdrop, highlighting the beauty of winter birdwatching and reminding us that life continues even in the harshest conditions.

Birdwatchers often turn to feeders and heated bird baths during winter, supplementing the scarce natural food sources during these months. Others bundle up and find joy in the frigid conditions. A friendly reminder that keeping your feeders and baths clean is essential to prevent the spread of disease. Setting up a feeder in your backyard can attract a wide variety of species, providing daily opportunities to observe their interactions and habits up close.

SPRING

As the snow melts and the days grow longer, spring ushers in one of the most exciting times for Great Lakes region birdwatchers: migration. This season is marked by the return of millions of birds from their wintering grounds in Central and South America. The Great Lakes are a critical stopover point for many species, making this region a hotspot for birdwatching throughout migration.

Warblers, sparrows, and shorebirds are among the many species that pass through the area. Their bright breeding plumage and lively songs add color and beautiful music to the landscape. Birdwatchers often hit the trails early to catch the peak of migration, when the trees and skies come alive with movement. I always say that the early bird nerd gets the bird, and the sense of anticipation during migration is genuinely exhilarating.

Key locations across the Great Lakes region become gathering points for birders hoping to catch glimpses of rare and diverse species. The unpredictability of migration—never knowing exactly what you'll see—adds excitement to every outing. Spring is also a time to sharpen your identification skills; the more species you can identify, the more rewarding the experience becomes. Hopefully, this book can help.

SUMMER

Summer in the Great Lakes region is nesting season. Birds are busy raising their young and defending their territories, and the air is filled with the sounds of fledglings begging for food. This season offers birdwatchers the chance to observe nesting behaviors, from the intricate construction of nests to the tireless feeding of chicks.

Species like Eastern Bluebird, American Goldfinch, and Great Blue Heron are familiar sights during summer. Wetlands and forests teem with life, providing ample opportunities to go birding. Birdwatchers may spend hours watching a single nest, witnessing the progress from eggs to fledglings to young birds taking their first flights.

Summer also presents the challenge of heat and thick foliage, making spotting birds more difficult, and early morning outings often yield the best results, as birds are most active during the cooler hours. The longer days allow for extended birdwatching sessions, and evening walks can be just as rewarding, with species like the American Woodcock displaying their unique courtship sky dances.

FALL

As summer fades into fall, the migration cycle begins again, in reverse. The Great Lakes region becomes a corridor for birds heading south to their wintering grounds. Fall migration tends to be more drawn out than spring migration, with different species moving through at other times.

This season offers birdwatchers a second chance to see many species that passed through in the spring and observe birds that may have been more elusive during their breeding season. Fall is also a time of change, as birds molt into their nonbreeding plumage, and the landscape transforms with the colors of autumn.

Key locations, such as hawk-watching sites and shorelines, become focal points for birdwatchers hoping to witness the migration spectacle. The fall migration is also an excellent time for citizen science projects, as birdwatchers can contribute valuable data on bird movements to conservation efforts.

THE YEAR-ROUND JOURNEY

Birdwatching in the Great Lakes region offers endless opportunities for discovery and connection with nature. Each season brings challenges and rewards, from winter's quiet beauty to migration's bustling energy. By attuning yourself to the year's rhythms, you can set yourself up for success as a birder and find bird joy and inspiration every season.

Whether you're a seasoned birder or just beginning your journey, this region provides a rich and diverse environment for birdwatching, where every outing promises something new. As you follow the cycle of the seasons, it's important to remember to respect the birds and their habitats. Avoid disturbing nesting birds, and always follow local regulations and guidelines. By doing so, you'll see how deeply interconnected the lives of birds are with the changing world around us.

Exploring Bird Habitat and Behavior in the Great Lakes Region

The Great Lakes region is not just a geographical expanse but a mosaic of habitats that support a wide array of bird species over the course of a year by providing food and/or nesting space.

LAKES, WETLANDS, AND MUDFLATS

Lakes and lakeshores are bustling hubs of bird activity, whether the Great Lakes themselves or one of the thousands of smaller lakes and ponds in the region. Associated wetlands and mudflats are key as well, providing vital resources for resident and migratory birds. Waterfowl like Mallards, Wood Ducks, and elegant Tundra Swans glide gracefully on the water's surface. Herons and egrets wade in the shallows, patiently stalking fish, while sandpipers probe the mud for food.

FORESTS AND WOODLANDS

Forests and woodlands are home to many songbirds, including warblers, thrushes, and woodpeckers. Warblers, with their bright plumage, flit among the branches during migration, adding a splash of color to the canopy. Woodpeckers drum on tree trunks, their rhythmic beats echoing through the woods. Owls, too, find prey and roosting space here.

GRASSLANDS AND FIELDS

Open grasslands and agricultural fields attract species like Dickcissel, Bobolink, and Eastern Bluebird. Raptors like Northern Harrier, Red-tailed Hawk, and American Kestrel soar above, scanning for prey.

URBAN AND SUBURBAN NEIGHBORHOODS

Birds can still thrive in urban and suburban settings. Trees, shrubs, and green spaces in neighborhoods provide habitat for Northern Cardinals, House Finches, hummingbirds, orioles, and Blue Jays. Planting native plants and shrubs and adding bird feeders and birdbaths help attract diverse species, allowing close looks at plumage and behavior.

PAY ATTENTION

Birdwatching means paying attention to birds' behavior, their seasonal or everyday habits, not just the color of their feathers. Behavior offers a glimpse into bird ecology, whether they're migrating, finding food, courting each other, or caring for young. As you explore the varied landscapes of the Great Lakes, take time to really watch birds and appreciate the beauty and complexity of their lives. Here are some general categories to keep in mind.

MIGRATION MARVELS The Great Lakes region is a crucial stopover for migratory birds traveling along the Mississippi and Atlantic Flyways. Witness flocks of waterfowl, shorebirds, and songbirds as they make their epic journeys northward in spring and southward in fall. Birders can experience the thrill of spotting rare species during migration and marvel at the resilience of these travelers.

FINDING FOOD Birds have to eat, and each species finds food in their own way. They might scratch for seeds on the ground, pick insects from foliage, dig into rotting wood for larvae, catch bugs midair, or silently sneak up on live prey. You've got to admire the talent needed to find food every day.

COURTSHIP DISPLAYS Springtime heralds the arrival of courtship displays among birds. From the intricate dances of Sandhill Cranes to the aerial acrobatics of American Woodcocks, these displays are a sight to behold. Male birds show off their plumage, songs, and behaviors to attract mates, adding a touch of romance to the birding season.

NESTING AND PARENTAL CARE Late spring and summer bring nesting season, a time of frenetic activity as birds build nests, incubate eggs, and raise young. Watch for parent birds tirelessly feeding and protecting their chicks with remarkable care. Observing these behaviors provides insights into the intricacies of bird family life.

Bird Families of the Great Lakes Region

A "family" is a group of closely related species that share physical characteristics and behaviors. Each family has a name in English and one in Latin, the language of scientific classification. Learning to recognize a bird family can help you narrow down the possible species when you are trying to identify an unfamiliar bird.

GEESE, SWANS & DUCKS (family Anatidae)

Look on any body of water and you'll probably find a duck or goose. These swimming birds have webbed feet, waterproof plumage, and a bill that matches their foraging method. The Mallard (*Anas platyrhynchos*), with its vibrant colors and dabbling behavior, is a familiar sight across the region.

LOONS (family Gaviidae)

These diving waterbirds are known for their haunting calls and are supremely adapted to fishing. Their webbed feet are set far back on their body for efficient underwater propulsion, but this placement makes them clumsy on land. The iconic Common Loon (*Gavia immer*) symbolizes northern lakes with its black-and-white plumage and eerie yodeling.

GREBES (family Podicipedidae)

These diving waterbirds have pointed bills, somewhat long necks, and lobed toes on legs set far back on the body. Another feature common to grebes is their lack of prominent tail feathers, giving their back end a fluffy look. The Pied-billed Grebe (*Podilymbus podiceps*) has a banded bill and is a characteristic species of lakes.

CORMORANTS (family Phalacrocoracidae)

These sleek, black, fish-eating birds have webbed feet, a long neck with a long, hooked bill, and excellent diving ability. The Double-crested Cormorant (*Nannopterum auritum*), with its yellow head feathers and fishing prowess, frequents lakes and rivers.

PELICANS (family Pelecanidae)

Pelicans are huge waterbirds recognized by their immense bills and throat pouches for scooping up fish. While not common in the Great Lakes region, the American White Pelican (*Pelecanus erythrorhynchos*) is occasionally spotted during migration.

BITTERNS, HERONS & EGRETS (family Ardeidae)

The graceful, long-necked, long-legged wading birds of this family are commonly found in or near wetlands, looking for prey to catch with daggerlike bills. One widespread member, the Great Blue Heron (*Ardea herodias*), is known for its skill in catching all manner of creatures.

RAILS & COOTS (family Rallidae)

These secretive, chicken-like marsh birds have slender bodies to slip through dense cattails and long toes ideal for walking on floating vegetation. The Virginia Rail (*Rallus limicola*), with its shy demeanor and loud calls during breeding season, is a challenging but rewarding find in wetlands.

CRANES (family Gruidae)

Cranes are graceful, long-legged, gregarious birds often associated with wetlands but sometimes spotted on suburban lawns. The Sandhill Crane (*Antigone canadensis*) is most common in marshy habitats, and their bugling calls can be heard across the Great Lakes region.

STILTS & AVOCETS (family Recurvirostridae)

These unusual long-necked, long-legged shorebirds (aka "waders") have distinctive plumage. The American Avocet's (*Recurvirostra americana*) upturned bill and rust-colored head and neck in the breeding season make it stand out from the crowd in wetlands and on mudflats.

PLOVERS (family Charadriidae)

These medium-sized, stout-billed shorebirds (aka "waders") frequent beaches and mudflats and pick food from the ground rather than probing. The Killdeer (*Charadrius vociferus*) is a familiar species with a distinctive *kill-deer* call and a broken-wing display to distract predators from its nest.

SANDPIPERS & SNIPES (family Scolopacidae)

This is a diverse group of shorebirds (aka "waders") with long bills perfect for probing mud and sand for invertebrates. The Least Sandpiper (*Calidris minutilla*), with its small size and fluttering flight over mudflats, is a common migrant along the Great Lakes shores.

GULLS & TERNS (family Laridae)

This family includes familiar white-and-gray gulls found near water and rarely far from land. They are adept at catching fish, but the bold ones will steal your lunch. The Ring-billed Gull (*Larus delawarensis*) has a black ring on its yellow bill and is a common sight near lakeshores and beaches.

TURKEYS, GROUSE & PHEASANTS (family Phasianidae)

These game birds usually stick to the ground, whether in grasslands, fields, or prairies, but may explode into flight if startled. The Wild Turkey (*Meleagris gallopavo*) is recognized by its impressive size and gobbling calls and symbolizes the region's woodlands and open fields.

NEW WORLD VULTURES (family Cathartidae)

Perhaps less glamorous than some other families, vultures play a crucial decomposition role in ecosystems, and their featherless heads won't become badly soiled when feeding on a dead animal. The Turkey Vulture (*Cathartes aura*) can be seen soaring effortlessly on wide wings, riding thermal currents in its search for carrion.

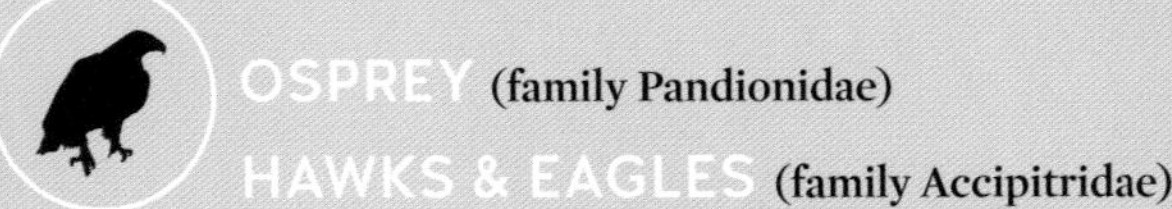

OSPREY (family Pandionidae)
HAWKS & EAGLES (family Accipitridae)

These are the majestic raptors known for keen eyesight, a short, hooked bill, needle-sharp talons, and impressive hunting skills. In the Great Lakes region, the Red-tailed Hawk (*Buteo jamaicensis*) is a common sight, soaring high above fields and woodlands in search of prey.

FALCONS (family Falconidae)

These smaller raptors are just as swift and deadly as the larger hawks and eagles, and they feed on everything from dragonflies to pigeons, often caught midair. The American Kestrel (*Falco sparverius*), with its colorful plumage and hovering hunting style, is a common sight in open habitats.

PIGEONS & DOVES (family Columbidae)

Both urban pigeons and peaceful doves are part of this family. Primarily fruit- and seed-eaters, they might forage on the ground or at bird feeders. The soft cooing of the subtly beautiful Mourning Dove (*Zenaida macroura*) is a common sound in suburban areas and open woodlands.

OWLS (family Strigidae)

Owls are nocturnal birds of prey known for cryptic coloration, keen hearing, silent flight, and powerful talons. The Barred Owl (*Strix varia*), with its haunting hoots and dark eyes, inhabits dense forests and wetlands, occasionally appearing near urban areas.

KINGFISHERS (family Alcedinidae)

These agile, large-headed, long-billed birds are always near water, where they dive for fish. The Belted Kingfisher (*Megaceryle alcyon*) has an attention-getting rattling call and striking blue-and-white plumage.

WOODPECKERS (family Picidae)

Woodpeckers have specialized chisel-like bills for digging into wood and extracting insects and larvae. Their short legs, zygodactyl toe arrangement (two toes pointing forward, two pointing back), and stiff tail feathers allow them to cling to tree trunks or other vertical surfaces. The Downy Woodpecker (*Dryobates pubescens*) is a common woodland resident, small in size with black-and-white plumage.

JAYS & CROWS (family Corvidae)

The opportunistic crows and jays are well known for their problem-solving abilities and have adapted to living near human habitation. They tend to be combinations of black and blue, sometimes have a crest, and use their sturdy bills to feed on everything from acorns to insects. The Blue Jay (*Cyanocitta cristata*), with its striking blue plumage and raucous calls, is a charismatic member of this family.

HUMMINGBIRDS (family Trochilidae)

At over 360 species, Trochilidae is the second most diverse bird family on Earth and comprises tiny, iridescent birds that hover in midair and feed on flower nectar. The iconic Ruby-throated Hummingbird (*Archilochus colubris*) does in fact have a radiant red throat and is a dazzling migrant that frequents gardens with nectar-rich flowers.

SWALLOWS (family Hirundinidae)

These agile aerial insectivores swoop and glide on the wing, gobbling down flying insects and appearing to never rest. The iridescent blue-green Tree Swallow (*Tachycineta bicolor*) is a charming summer resident and will take up residence in the right kind of bird house.

CHICKADEES & TITMICE (family Paridae)

These are small, active year-round birds with tiny bills that forage in trees and shrubs looking for insects and seeds. The Black-capped Chickadee (*Poecile atricapillus*), with its *chick-a-dee-dee-dee* call and bold behavior, is a year-round companion in woodlands.

NUTHATCHES (family Sittidae)
CREEPERS (family Certhiidae)

Nuthatches are small, acrobatic birds that climb down tree trunks headfirst and hang upside down from branches as they forage for seeds or insects. The White-breasted Nuthatch (*Sitta carolinensis*) has a habit of wedging seeds in bark crevices. The Brown Creeper (*Certhia americana*) climbs up trunks in a spiral, looking for insects.

WRENS (family Troglodytidae)

Wrens are small, brown, energetic birds with big voices. Whether in the suburbs or a wetland, a wren will be heard before it's seen. With a loud trilling song, the House Wren (*Troglodytes aedon*) is a yearly summer migrant to woodlands and neighborhoods.

VIREOS (family Vireonidae)

These small, insectivorous spring migrants have muted green and gray plumage, a thick, subtly hooked bill, and melodic songs. The aptly named Red-eyed Vireo (*Vireo olivaceus*) sings continuously during breeding season and is a characteristic species of deciduous forests and woodlands.

KINGLETS (family Regulidae)

This family of tiny, active birds includes one of the smallest in North America, the Golden-crowned Kinglet. Kinglets are generally greenish to greenish yellow and forage for insects in trees and shrubs using their minuscule bills. The crown of the Ruby-crowned Kinglet (*Regulus calendula*) is usually concealed, but the bird's rapid movements will catch your eye.

FLYCATCHERS (family Tyrannidae)

This is the most diverse bird family on Earth, with over 400 species. It includes small and medium-sized insect-eaters known for their sallying flights to catch prey in midair. The tail-wagging behavior and *phoebe* call of the Eastern Phoebe (*Sayornis phoebe*) is a common sight and sound near water and open habitats in summer.

THRUSHES (family Turdidae)

This family of fruit-, insect-, and worm-eating birds encompasses the spotted thrushes, the bluebirds, and the familiar American Robin (*Turdus migratorius*). Many have melodic songs or colorful feathers. The robin has both, with its orange breast and cheerful song, and is a much-loved herald of spring and summer across urban and rural landscapes.

WAXWINGS (family Bombycillidae)

The three species of waxwings worldwide are known for their sleek appearance, crested heads, color-accented feathers, and penchant for berries. The Cedar Waxwing (*Bombycilla cedrorum*), with its yellow-tipped tail, is the only strictly North American waxwing and is a delightful visitor to fruiting trees.

CATBIRDS, THRASHERS & MOCKINGBIRDS (family Mimidae)

These vocal mimics develop impressive song repertoires over their lives. They're often solitary and secretive, more often heard than seen, and they forage by scratching through leaf litter or picking berries from shrubs. The Northern Mockingbird (*Mimus polyglottos*) is known to copy other bird species and urban sounds and adds musical diversity to suburban areas.

STARLINGS (family Sturnidae)

Starlings are found worldwide and are highly vocal songbirds known for their gregarious behavior. While not originally native to North America, the European Starling (*Sturnus vulgaris*) has become a familiar sight in urban and suburban areas. It sports iridescent plumage and mimics a variety of sounds.

GNATCATCHERS (family Polioptilidae)

These are tiny, busy, long-tailed gray birds that glean insects from foliage. The Blue-gray Gnatcatcher (*Polioptila caerulea*) can be found in wooded areas, fanning its white-edged tail, and flitting about energetically.

FINCHES (family Fringillidae)

These seedeating birds are equipped with conical bills just right for extracting and cracking seeds. Finches usually feed above the ground, on seed heads, in trees, or at bird feeders. The unmistakable male American Goldfinch (*Spinus tristis*) adds a splash of color with its lemon-yellow breeding plumage and melodious song.

SPARROWS & TOWHEES (family Passerellidae)

Sparrows are small, often brown, seedeating birds with conical bills that occupy diverse habitats. They are most comfortable feeding on the ground, scratching for seeds. The Song Sparrow (*Melospiza melodia*), with its streaked plumage and melodious trill, is a common inhabitant of marshes and grasslands.

OLD WORLD SPARROWS (family Passeridae)

While our native sparrows are known as New World sparrows, two superficially similar species introduced to North America are known as Old World sparrows and are in a different family. The House Sparrow (*Passer domesticus*) is habituated to living with humans and can be found anywhere there is spilled grain or crumbs, whether a farm or an outdoor café.

SNOW BUNTING (family Calcariidae)

This sparrow-like bird is adapted to colder climates. In winter, Snow Buntings (*Plectrophenax nivalis*) grace snowy landscapes, their black-and-white plumage contrasting with the frosty backdrop.

TANAGERS, CARDINALS, GROSBEAKS, BUNTINGS & DICKCISSELS (family Cardinalidae)

This diverse family boasts vibrant songbirds that add splashes of color to woodlands or grasslands. Most are spring and summer migrants, but the familiar Northern Cardinal (*Cardinalis cardinalis*) is a beloved resident with bright red plumage, a prominent crest, and a melodious song.

BLACKBIRDS (family Icteridae)

This diverse family includes blackbirds and grackles but also orioles and meadowlarks, all with distinctive vocalizations and behaviors. Most have slender, pointed bills and long tails, with some exceptions, and are found in open areas or woodlands, depending on species. The Red-winged Blackbird (*Agelaius phoeniceus*) thrives in marshes and grasslands and is recognized by its red shoulder patches and territorial displays.

WOOD-WARBLERS (family Parulidae)

The wood-warblers add vibrancy to spring and fall migration, but with over 30 species passing through the region, they can also overwhelm birders. Warblers are small, slim, insectivorous birds with varying amounts of yellow on their bodies. They're found in most woodlands but can be inconspicuous despite their bright breeding plumage. Warblers are typically very active while searching for food, constantly flitting and darting through foliage. The Yellow Warbler (*Setophaga petechia*), with its bright yellow plumage and sweet song, epitomizes the beauty of warbler species.

Warbler plumage differs between male and female, and between spring and fall. This book illustrates spring breeding males, with the exception of the Wilson's Warber, but plenty of resources exist for your further warbler study.

FIELD GUIDE

Adult.
Black head and neck with white chinstrap. Brown body.

Juvenile.
Yellowish and downy.

CANADA GOOSE

Branta canadensis

Brown body with long black neck and black head, highlighted by striking white chinstrap and cheek patches. In flight, tail displays a beautiful pattern of black and white.

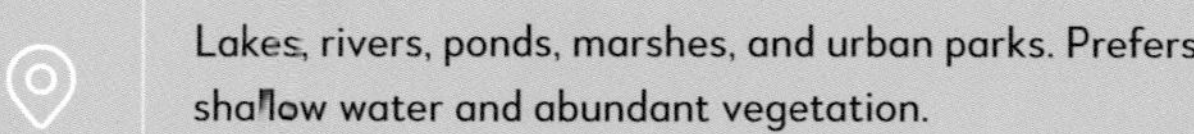

Lakes, rivers, ponds, marshes, and urban parks. Prefers shallow water and abundant vegetation.

Highly adaptable and social. Feeds on grasses, aquatic plants, grains, seeds, and agricultural crops. Honking used for communication within flocks and during flight to maintain group cohesion. Forms monogamous pairs and builds large nests near water bodies.

These iconic birds are famous for their impressive migrations, often traveling 1500 miles in a single day! Their long-distance journeys demonstrate incredible endurance and navigational skills.

LENGTH: 29–43" / WINGSPAN: 50–66"

Adult.
Huge white bird with huge black bill. Black facial skin between bill and eye. Plumage may be stained by tannins in the water.

TRUMPETER SWAN

Cygnus buccinator

Impressive size, pure white plumage, long straight neck, and huge black bill. Unlike Tundra Swan, it lacks yellow spot around the eye.

Lakes, marshes, and rivers with ample vegetation for feeding and nesting.

Feeds mainly on plants but also small fish, aquatic insects, and fish eggs. Elaborate courtship displays involve synchronized swimming, head bobbing, and vocalization. During breeding season, they form pairs or small family groups and construct large nests near water. Populations vary, with some year-round residents and others seasonal migrants.

The Trumpeter Swan holds the title of largest species of swan and is one of the largest birds on the planet.

LENGTH: 54–62" / WINGSPAN: 79–96"

Adult.
Huge white bird. Spot of yellow at base of black bill.

TUNDRA SWAN

Cygnus columbianus

Large white swan, long straight neck, large black bill with yellow spot at base. Smaller overall than Trumpeter Swan and distinguished by presence of yellow spot at base of bill.

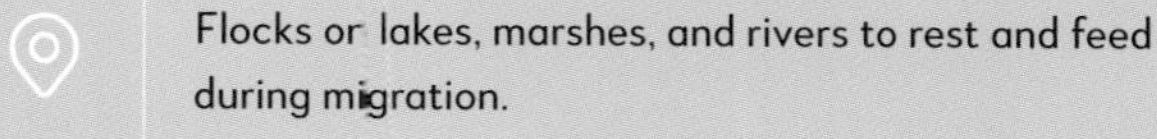

Flocks on lakes, marshes, and rivers to rest and feed during migration.

Feeds mainly on plant matter but also consumes mollusks and arthropods unearthed from the mud. Migrates annually between Arctic tundra breeding grounds and more temperate wintering grounds. Forms monogamous pairs during breeding season, with both parents incubating and caring for the cygnets until they can fend for themselves.

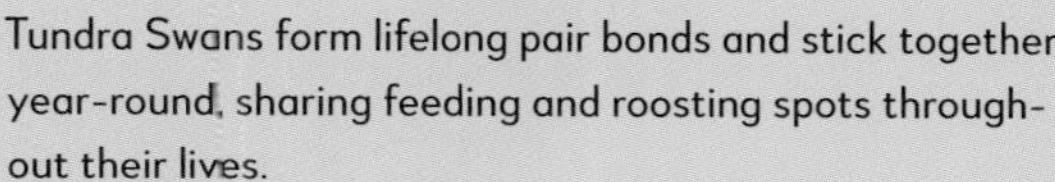

Tundra Swans form lifelong pair bonds and stick together year-round, sharing feeding and roosting spots throughout their lives.

LENGTH: 47–57" / WINGSPAN: 66"

Male.
Slate-blue head, bold white crescent.

Female.
Overall brown, but blue patch in wing.

BLUE-WINGED TEAL

Spatula discors

Small dabbling duck with long bill, slate-blue head and neck, cinnamon-brown spotted body. Male has bold white crescent on face. Female mottled brown and tan. Both have blue wing patches bordered by white, visible in flight.

Marshes, ponds, lakeshores, and flooded fields with ample aquatic vegetation. Prefers shallow freshwater for food and suitable nesting sites.

Highly social dabbling ducks, often in flocks during migration and in winter. Feed by upending in the water to reach aquatic plants, seeds, and invertebrates. Agile flyers, capable of swift and twisting flight patterns.

Unlike other ducks, Blue-winged Teals migrate north in late spring and south in early fall.

LENGTH: 14–16" / WINGSPAN: 22–24"

Breeding male.
Glossy green head with yellow bill. White ring around neck.

Breeding female.
Mottled brown body. Orange bill with varying black. Dark eyeline.

MALLARD

Anas platyrhynchos

Medium-sized duck exhibiting striking differences between males and females. Male has vibrant green head, white neck ring, and chestnut-brown chest on grayish body. Female has mottled brown plumage, orange-and-black bill, and excellent camouflage.

Lakes, ponds, rivers, marshes, and urban parks. Prefers shallow water, emergent vegetation, and nearby grassy or wooded areas.

Feeds on aquatic plants, seeds, grains, and small invertebrates by dabbling at the water's surface or upending to access food below the surface. Highly adaptable and thrives in urban and suburban environments near water.

Mallards can take flight straight up from the water, while larger birds must run along the surface before lifting off.

LENGTH: 19–25" / WINGSPAN: 32–37"

Breeding male.
Gray body with vertical white stripe at shoulder. Chestnut head with iridescent green patch.

GREEN-WINGED TEAL

Anas crecca

Smallest teal species. Male's head chestnut with iridescent green streak behind eye; back speckled gray-brown. Vertical white stripe at shoulder. In flight, green patch shows in wings. Female mottled brown with dark eyeline.

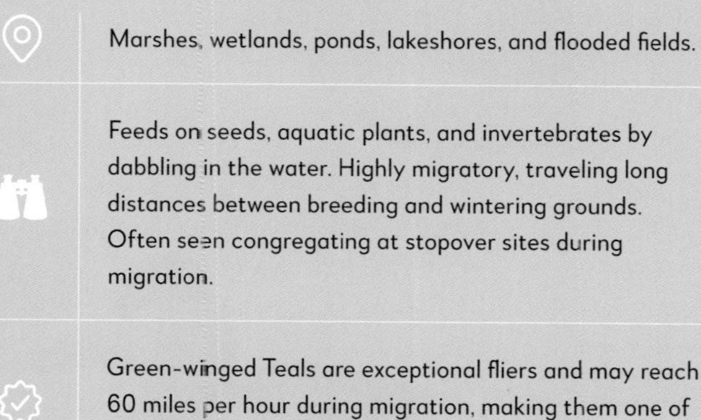

Marshes, wetlands, ponds, lakeshores, and flooded fields.

Feeds on seeds, aquatic plants, and invertebrates by dabbling in the water. Highly migratory, traveling long distances between breeding and wintering grounds. Often seen congregating at stopover sites during migration.

Green-winged Teals are exceptional fliers and may reach 60 miles per hour during migration, making them one of the fastest duck species in the world.

LENGTH: 12–15" / WINGSPAN: 20–23"

Male.
Dark head and back end with white middle. Head slightly peaked. Yellow eye.

Female.
All-over brown, with varying white at base of bill. Slightly peaked head.

LESSER SCAUP

Aythya affinis

Medium-sized diving duck. Male has peaked black head and black neck, white flanks, and light gray back. Bill blue-gray with black tip. Female also has peaked head but all over brown with white at base of bill.

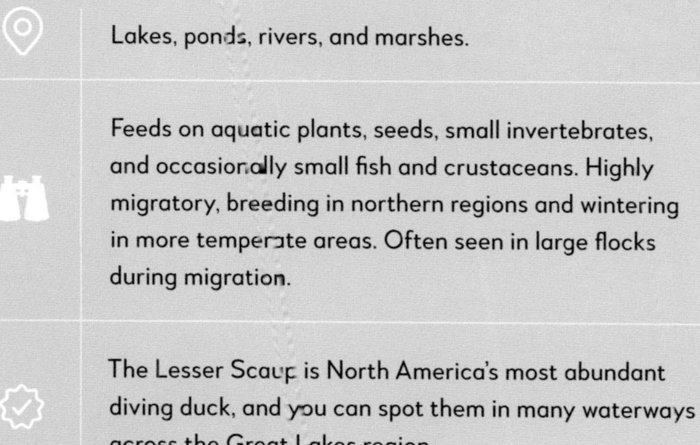

Lakes, ponds, rivers, and marshes.

Feeds on aquatic plants, seeds, small invertebrates, and occasionally small fish and crustaceans. Highly migratory, breeding in northern regions and wintering in more temperate areas. Often seen in large flocks during migration.

The Lesser Scaup is North America's most abundant diving duck, and you can spot them in many waterways across the Great Lakes region.

LENGTH: 15–18" / WINGSPAN: 26–30"

Breeding male.
Black face and neck, rest of head white. Bright white belly.

Breeding female.
Dark all over with white patch behind and below eye.

BUFFLEHEAD

Bucephala albeola

Small diving duck with compact body; large, rounded head; stubby bill; and striking black-and-white plumage. Up close, males have iridescent purple and green sheen to black head.

Shallow lakes, ponds, marshes, and rivers with abundant vegetation.

Skilled divers that forage underwater for aquatic invertebrates, small fish, and plant matter. Males perform elaborate courtship displays in the water and chatter to attract females.

The Bufflehead is the smallest diving duck in North America, making it a pint-sized powerhouse of aquatic prowess.

LENGTH: 12–15" / WINGSPAN: 21"

Male's bright white crest and yellow eye are distinctive. Female is all-over brown with a cinnamon-colored crest resembling a fan brush.

HOODED MERGANSER

Lophodytes cucullatus

Male distinguished by eye-catching black, white, and brown body; large, fan-shaped white crest with black border; and yellow eye. Female brownish all over with cinnamon-colored crest, subtle yet elegant look. Both have long, narrow bill.

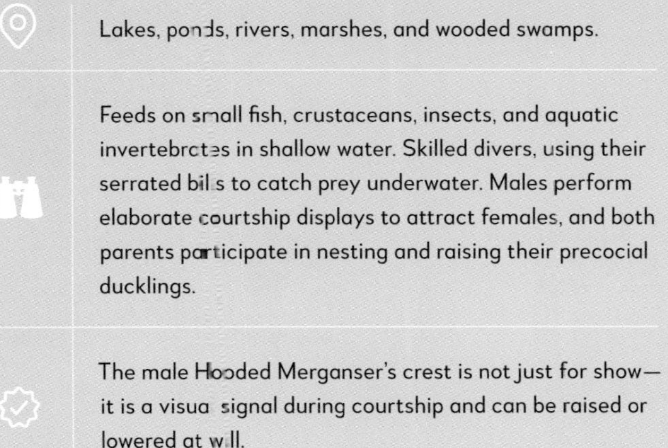

Lakes, ponds, rivers, marshes, and wooded swamps.

Feeds on small fish, crustaceans, insects, and aquatic invertebrates in shallow water. Skilled divers, using their serrated bills to catch prey underwater. Males perform elaborate courtship displays to attract females, and both parents participate in nesting and raising their precocial ducklings.

The male Hooded Merganser's crest is not just for show—it is a visual signal during courtship and can be raised or lowered at will.

LENGTH: 15–19" / WINGSPAN: 23–26"

Breeding male.
Narrow red bill on dark green head. Mostly white body with black back.

Breeding female.
Narrow pinkish bill on rust-colored head with ragged crest. Gray body.

COMMON MERGANSER

Mergus merganser

Large, striking diving duck with sleek, elongated body and long, narrow red bill. Male's head dark green; body black and white. Female's head rust colored with short crest; body light gray.

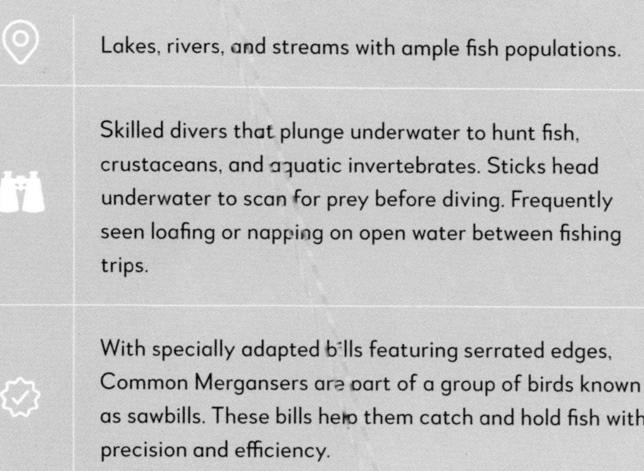

Lakes, rivers, and streams with ample fish populations.

Skilled divers that plunge underwater to hunt fish, crustaceans, and aquatic invertebrates. Sticks head underwater to scan for prey before diving. Frequently seen loafing or napping on open water between fishing trips.

With specially adapted bills featuring serrated edges, Common Mergansers are part of a group of birds known as sawbills. These bills help them catch and hold fish with precision and efficiency.

LENGTH: 21–27" / WINGSPAN: 33"

Adult.
Blocky head, daggerlike bill, red eye. Checkerboard back.

Adult.
Zebra-stripe necklace. White belly.

COMMON LOON

Gavia immer

Large waterbird with daggerlike bill. Sleek black head and neck contrast with white belly and underparts. Back and wings display checkerboard pattern during breeding season, which turns to uniform gray in nonbreeding.

Lakes with clear, deep water and sufficient fish populations.

Excellent underwater hunters, using powerful legs and feet set far back on the body to propel themselves in pursuit of fish and other aquatic prey. Builds a floating nest of aquatic vegetation. Haunting calls and wails are used for territorial defense, communication, and coordination during breeding and migration.

Because their legs and feet are so far back, a Common Loon cannot take off straight up and must run on the water while flapping its wings to gain enough speed for flight.

LENGTH: 26–35" / WINGSPAN: 40–51"

Breeding male.
Brown body. White eyering.
Gray bill with black ring.

PIED-BILLED GREBE

Podilymbus podiceps

Compact waterbird with short neck and thick bill. Brownish-gray plumage with black band across the bill during breeding season—the "pied" bill. Band fades to pale color during non-breeding season.

Marshes, ponds, lakes, and slow-moving rivers with dense emergent vegetation.

Feeds on small fish, crustaceans, insects, and other aquatic invertebrates. Skilled diver with lobed feet to propel itself underwater in search of prey. Secretive, often sinking underwater to avoid detection. Builds floating nest in dense vegetation.

The Pied-billed Grebe may look like a duck, but its lobed toes, chicken-like bill, and lack of tail feathers put it in the grebe family.

LENGTH: 11–15" / WINGSPAN: 17–24"

Adult.
Dark body. Orange hooked bill.

Juvenile.
Pale body. Orange hooked bill.

DOUBLE-CRESTED CORMORANT

Nannopterum auritum

Large dark brown or glossy black waterbird. Slender neck supports long, hooked bill, perfect for catching fish underwater. Vibrant orange-yellow skin around bill and chin.

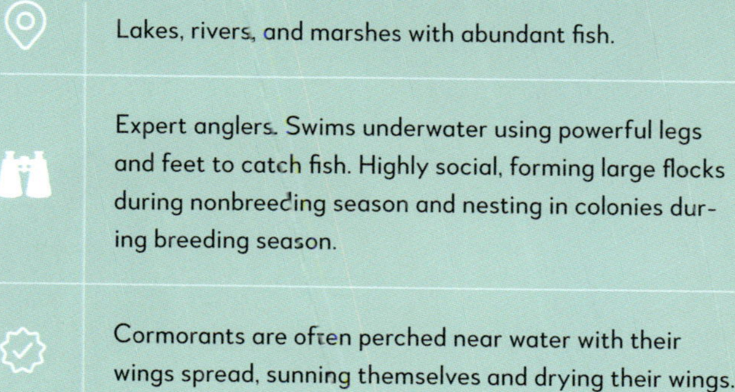

Lakes, rivers, and marshes with abundant fish.

Expert anglers. Swims underwater using powerful legs and feet to catch fish. Highly social, forming large flocks during nonbreeding season and nesting in colonies during breeding season.

Cormorants are often perched near water with their wings spread, sunning themselves and drying their wings.

LENGTH: 27–35" / WINGSPAN: 44–48"

Adult.
Huge white bird with black flight feathers, massive orange bill.

Adults preening.
Pelicans are highly social.

AMERICAN WHITE PELICAN

Pelecanus erythrorhynchos

Huge, majestic, sociable bird with up to 9-foot wingspan. Long yellow bill with throat pouch, and white plumage accented by black flight feathers. During breeding season, adults develop conspicuous bump on bill.

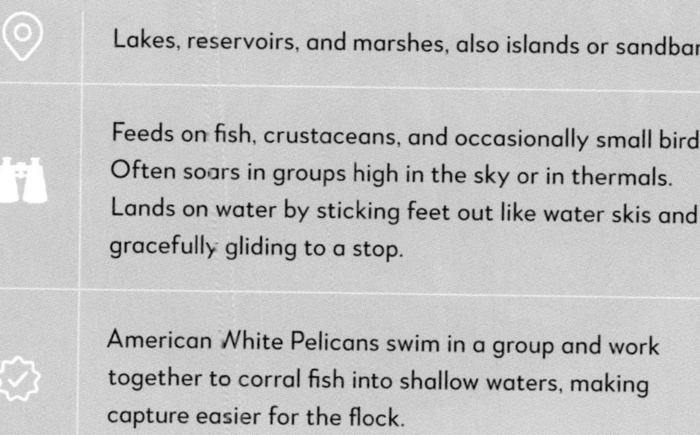

Lakes, reservoirs, and marshes, also islands or sandbars.

Feeds on fish, crustaceans, and occasionally small birds. Often soars in groups high in the sky or in thermals. Lands on water by sticking feet out like water skis and gracefully gliding to a stop.

American White Pelicans swim in a group and work together to corral fish into shallow waters, making capture easier for the flock.

LENGTH: 50–65" / WINGSPAN: 96–114"

Juvenile.
Striped chest. Buff, rust, and white head and back.

LEAST BITTERN

Ixobrychus exilis

Very small heron. Slim body with brown and buff-colored feathers. Long, thin yellow bill. Male's back and crown dark green, with brown-and-white striped chest. Female's back and crown brown.

Secluded marshes and wetlands with dense emergent vegetation such as cattails.

Feeds on small fish, amphibians, insects, and other invertebrates. Solitary. Crepuscular (active dusk and dawn) and nocturnal but sometimes active by day. Skilled hunter that patiently stalks and then strikes with long bill.

When caught off guard, Least Bitterns freeze in place, bill pointed upward, mimicking the swaying motions of nearby reeds and blending into their marshy surroundings.

LENGTH: 11–14" / WINGSPAN: 16–18"

Adult.
Cryptic brown and buff plumage. Yellow bill. Dark mustache, white throat.

AMERICAN BITTERN

Botaurus lentiginosus

Medium-sized heron with stunningly cryptic plumage of mottled brown and buff tones. Feathers intricately streaked and barred, blending seamlessly with marshy surroundings.

Marshes, bogs, and swamps with dense, obscuring vegetation.

Adept at concealment, often stands motionless, neck extended upward to mimic surrounding reeds and grasses. A true master of camouflage.

During the breeding season, male American Bitterns emit resonant *oonk-a-lunk* or *pump-er-lunk* calls from inflated throat sacs. These calls reverberate across the wetlands, alluring potential mates.

LENGTH: 23–33" / WINGSPAN: 36"

Adult.
Stocky gray-blue body, yellow legs. Black crown and back. Red eye. Sturdy dark bill.

Juvenile.
Stocky striped body, yellow legs. Sturdy bill. Compare with bitterns and juvenile Green Heron.

BLACK-CROWNED NIGHT HERON

Nycticorax nycticorax

Medium-sized stocky heron with short neck and legs and broad, rounded head with red eyes. Plumage is predominantly grayish blue, with dark crown and back.

Marshes, wetlands, swamps, and shores of lakes, rivers, and ponds. Prefers areas with dense vegetation.

Crepuscular (active dusk and dawn) and nocturnal. Roosts in vegetation by day; feeds on fish, crustaceans, amphibians, insects, and small mammals by night. Employs patient and stealthy hunting technique and uses sharp bill to spear or grasp prey in shallow water.

Adult Black-crowned Night Herons don't distinguish between their chicks and those of other herons, caring for any chick that appears in their nest.

LENGTH: 22–26" / WINGSPAN: 45–46"

Adult.
Reddish sides of neck, striped chest, greenish crown and back. Yellowish legs.

Juvenile.
More striped all over. Compare with bitterns and juvenile Black-crowned Night Heron.

GREEN HERON

Butorides virescens

Small wading bird with vibrant greenish plumage, yellowish legs, dark greenish-blue back, chestnut-colored neck, and greenish-black crest.

Marshes, wetlands, ponds, streams, and wooded swamps with suitable overhanging perches for hunting.

Feeds on small fish, frogs, crustaceans, and insects. Patient hunter: stands motionless on shore or on branches overhanging the water, waiting to capture prey with sharp bill.

The Green Heron entices prey by dropping feathers, small sticks, or insects into the water, a clever and innovative approach to catching a meal.

LENGTH: 16–18" / WINGSPAN: 25–26"

Adult.
All-white body. Slim black bill.
Black legs with yellow feet.

SNOWY EGRET

Egretta thula

Medium-sized wading bird with all-white plumage, slim black bill, and long, slender black legs with bright yellow feet. During the breeding season, adults develop showy plumes on head, neck, and back.

Marshes wetlands, estuaries, and along the shores of lakes and rivers.

Wades in shallow water, stirs up sediment with their long legs, and flushes out small fish and aquatic invertebrates. Spears prey with sharp bill in a lightning-fast strike. Graceful and deliberate, often seen stalking prey or standing perfectly still for extended periods, waiting to strike.

At the beginning of the twentieth century, hunters had nearly destroyed Snowy Egrets for their breeding plumes, highly sought-after adornments for women's hats.

LENGTH: 22–26" / WINGSPAN: 39"

Adult.
Daggerlike yellow bill. All-white body. Black legs and feet.

GREAT EGRET

Ardea alba

Tall, elegant, all-white wading bird with long, slender neck, daggerlike yellow bill, and black legs and feet.

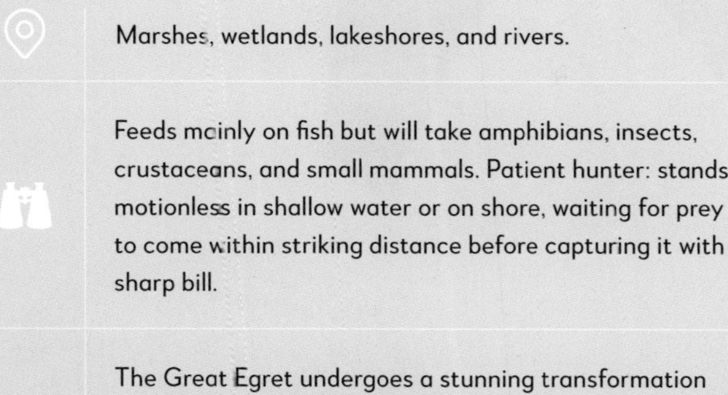

Marshes, wetlands, lakeshores, and rivers.

Feeds mainly on fish but will take amphibians, insects, crustaceans, and small mammals. Patient hunter: stands motionless in shallow water or on shore, waiting for prey to come within striking distance before capturing it with sharp bill.

The Great Egret undergoes a stunning transformation during the breeding season. Its facial patch turns vibrant neon-green and elegant long plumes adorn its back, adding a touch of glamour to this already majestic bird.

LENGTH: 37–40" / WINGSPAN: 51–57"

Adult.
In flight, the neck is tucked in.

Adult.
Daggerlike yellow bill, white face, gray cap. Long gray neck atop blue-gray body and long legs.

GREAT BLUE HERON

Ardea herodias

Tall wading bird with long, S-shaped neck and daggerlike bill. Gray-blue body, white head and neck. Black plumes extend from crown during breeding season.

Marshes, wetlands, rivers, and lakeshores.

Feeds mainly on fish but will take amphibians, insects, crustaceans, and small mammals. Patient hunter: will stand motionless in shallow water or on shore, waiting to strike at passing prey with sharp bill. Forms breeding colonies called heronries or rookeries near water and builds large stick nests lined with softer materials.

Great Blue Herons can stand motionless for extended periods while hunting. This patient approach enables them to blend into their surroundings, making them highly effective predators day or night.

LENGTH: 38–53" / WINGSPAN: 65–79"

Adult.
Slender orange bill. Light eyebrow. Patterned plumage; stubby tail.

VIRGINIA RAIL

Rallus limicola

Small, secretive chicken-like bird with long, slender orange bill; short, upright tail; and reddish-brown body marked with streaks of black and white.

Marshes, wet meadows, and reed beds. Prefers emergent vegetation and tall grasses.

Camouflages itself within thick vegetation to evade detection, but grunting or pig-like squeals give it away. Effortlessly navigates through dense plants, occasionally coming out into the open. Uses long toes to tread lightly on floating plants while hunting for insects, snails, and small crustaceans with its long bill.

Virginia Rails have developed ingenious survival tactics, crafting floating nests hidden in dense marsh vegetation and even creating decoy nests to outsmart marsh predators.

LENGTH: 7–10" / WINGSPAN: 12–15"

Adult.
Yellow bill. Black mask.
Mottled brown and gray body.

SORA

Porzana carolina

Small marsh bird in mottled brown and gray plumage. Gray face accented by black mask and short yellow bill. Yellow legs and upright tail give it chicken-like appearance.

Marshes, wetlands, and shallow ponds. Prefers dense vegetation.

Secretive and solitary. Flicks tail while foraging for seeds, aquatic plants, insects, and other invertebrates. Can run swiftly or swim when necessary.

Despite being small and secretive, these intriguing birds are skilled vocalists renowned for loud, whinnying calls.

LENGTH: 7–9" / WINGSPAN: 12"

Adult.
Dark body and head. White face and bill. Lobed toes.

AMERICAN COOT

Fulica americana

Compact, round blackish-gray body with small head. White bill and prominent frontal shield, reminiscent of bald forehead, distinguish it from other bird species.

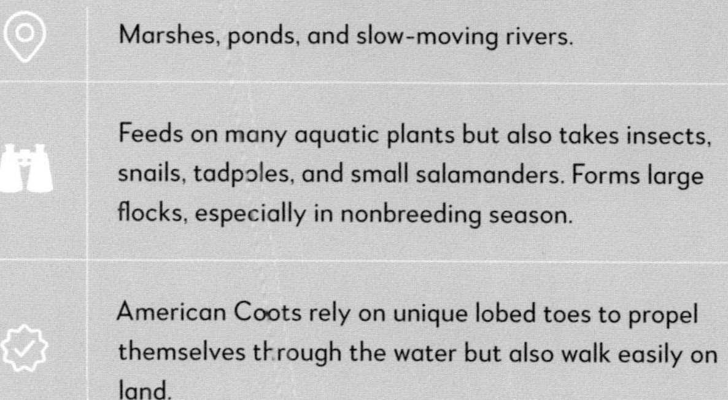

Marshes, ponds, and slow-moving rivers.

Feeds on many aquatic plants but also takes insects, snails, tadpoles, and small salamanders. Forms large flocks, especially in nonbreeding season.

American Coots rely on unique lobed toes to propel themselves through the water but also walk easily on land.

LENGTH: 15–16" / WINGSPAN: 23–25"

Adult.
Red cap. Gray and white head and neck. Long bill.

Adult.
Rust-colored feathers in summer.

SANDHILL CRANE

Antigone canadensis

Tall, gray, long-necked bird on long black legs. Red crown and white cheeks, daggerlike black bill. Adults 3–4 feet tall with 5–7 foot wingspan. In flight, neck and legs stick straight out.

Marshes, wet meadows, and grasslands, often near lakes, rivers, and wetlands. Occasionally in parks or suburban neighborhoods near suitable habitat.

Feeds on seeds, grains, roots, tubers, berries, invertebrates, and small vertebrates. Highly social. Known for loud, bugling calls. Performs elaborate courtship dances with intricate movements and displays. Pairs form lifelong bonds and demonstrate strong family ties during breeding and nesting.

In the early twentieth century, Sandhill Cranes faced extinction from habitat loss and overhunting. However, their populations have rebounded, a testament to the dedicated conservation efforts led by the late conservationist Aldo Leopold in Wisconsin.

LENGTH: 36–48" / WINGSPAN: 60–84"

Adult.
White body. Red face and crown.

WHOOPING CRANE

Grus americana

Extra-large, long-necked, snowy white bird around 5 feet tall with wingspan to about 7 feet. Black wingtips, red face and crown, long black legs, and long, straight black bill.

Wetlands, marshes, and shallow lakes, ponds, and rivers. Requires abundant vegetation for nesting.

Feeds on aquatic invertebrates, small fish, amphibians, and occasionally small mammals. Performs elegant courtship displays involving dances, bowing, and calling to strengthen pair bonds and establish territories.

Whooping Crane populations have drastically declined because of habitat loss, hunting, and human-induced factors. Conservation efforts aim to stabilize and increase their critically low populations, making them a symbol of conservation efforts in the Great Lakes region.

LENGTH: 59" / WINGSPAN: 90"

Adult.
In flight, long pink legs stick out.

Adult.
Thin bill. Black head and back. White throat and belly. Long pink legs.

BLACK-NECKED STILT

Himantopus mexicanus

Elegant wader with slender black-and-white body, ridiculously long pink legs, and very thin black bill.

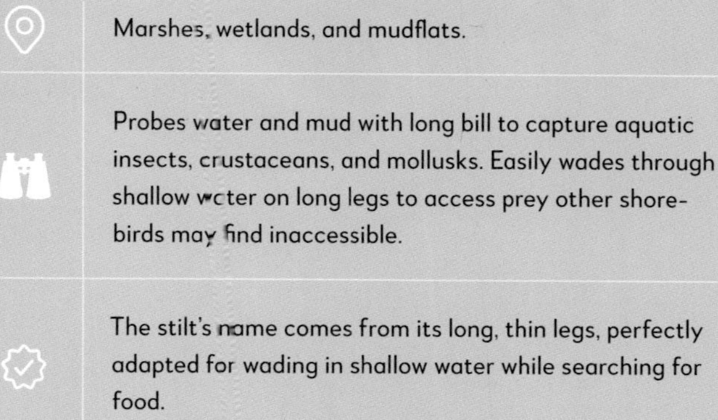

Marshes, wetlands, and mudflats.

Probes water and mud with long bill to capture aquatic insects, crustaceans, and mollusks. Easily wades through shallow water on long legs to access prey other shore-birds may find inaccessible.

The stilt's name comes from its long, thin legs, perfectly adapted for wading in shallow water while searching for food.

LENGTH: 13–15" / WINGSPAN: 28–29"

Breeding.
Upturned bill, reddish head and neck, black-and-white wings, long legs.

AMERICAN AVOCET

Recurvirostra americana

Graceful wader with slender, upturned bill and elegant black-and-white plumage. Head and neck cinnamon-colored in breeding plumage.

Shallow lakes, marshes, and mudflats rich in invertebrate prey.

Elegant and efficient foraging behavior: sweeps bill through the water, capturing small crustaceans, insects, and aquatic plants. Melodious whistled call adds a musical element to the scenery.

American Avocets experience a surprising change in plumage between the breeding and nonbreeding seasons. During the breeding season, their feathers turn a rusty orange; in the nonbreeding season, these areas fade to pale gray or white.

LENGTH: 16–18" / WINGSPAN: 28"

Breeding.
Stout bill. Round head. Black-and-white back and wings, with a black face, chest, and belly.

Nonbreeding.
Mottled brown-and-white plumage.

BLACK-BELLIED PLOVER

Pluvialis squatarola

Large shorebird with stout black bill. In breeding plumage, black-and-white back and wings; black face, chest, and belly; white crown and back of neck. Mottled brown-and-white non-breeding plumage.

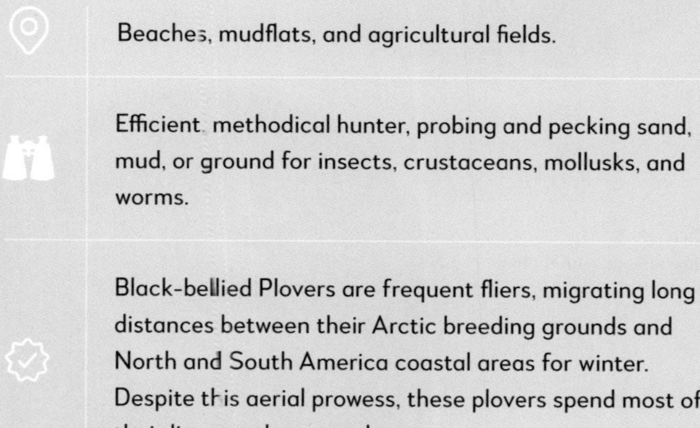

Beaches, mudflats, and agricultural fields.

Efficient, methodical hunter, probing and pecking sand, mud, or ground for insects, crustaceans, mollusks, and worms.

Black-bellied Plovers are frequent fliers, migrating long distances between their Arctic breeding grounds and North and South America coastal areas for winter. Despite this aerial prowess, these plovers spend most of their lives on the ground.

LENGTH: 11" / WINGSPAN: 23"

Adult.
Two black bands across chest. White forehead, white eyebrow, red eyerings.

Juvenile.
Puffy down mimics adult plumage pattern.

KILLDEER

Charadrius vociferus

Easily recognized by medium size, brown back, white underparts, two black bands across chest. Large eyes with red outline.

Fields, pastures, grasslands, and shorelines. Prefers short vegetation, gravel, or sandy substrates.

Feeds on insects, earthworms, and small invertebrates. Also consumes seeds and plant matter, particularly during nonbreeding season.

Killdeers are known for their broken-wing display, which distracts predators from nests or chicks and showcases their clever and protective behavior.

LENGTH: 7–11" / WINGSPAN: 18"

Breeding.
Tiny orange bill tipped with black. White at base of bill. Brown back, white belly. Single black chest band.

SEMIPALMATED PLOVER

Charadrius semipalmatus

Small, sleek shorebird with mottled brown plumage, black chest band, white belly, black crown, orange bill with black tip, and yellow legs.

Mudflats, agricultural ponds, sod farms, beaches, and shorelines.

Employs fast, darting movements while searching for insects, crustaceans, and other small invertebrates. Uses feet to stir up prey in shallow water, a memorable hunting technique.

The Semipalmated Plover gets its name from its partially webbed feet. This feature helps them navigate mud or sand while foraging.

LENGTH: 6–7" / WINGSPAN: 18–19"

Nonbreeding.
Long two-tone bill.
Grayish-brown mottled plumage.

HUDSONIAN GODWIT

Limosa haemastica

Medium-sized sleek, slender shorebird with long legs and slightly upturned bicolored bill. Mottled dark brown and gray plumage. Rusty belly on breeding adults.

Mudflats, estuaries, marshes, shorelines, wetlands, and flooded fields. Primarily spotted during migration while traveling between Arctic breeding grounds and South American wintering areas.

Feeds on aquatic insects, mollusks, and crustaceans. Uses long bill to probe mud or shallow water during migration stopovers.

The Hudsonian Godwit flies nonstop for thousands of miles over the open ocean between its breeding and wintering grounds.

LENGTH: 14–16" / WINGSPAN: 27–29"

Breeding.
Black belly. Warm brown back. Slightly drooping bill.

DUNLIN

Calidris alpina

Small shorebird with mottled brown plumage, black streaks. Conspicuous black belly patch in breeding season. Pale grayish-white underside in nonbreeding plumage. Distinctive bill curves downward just slightly.

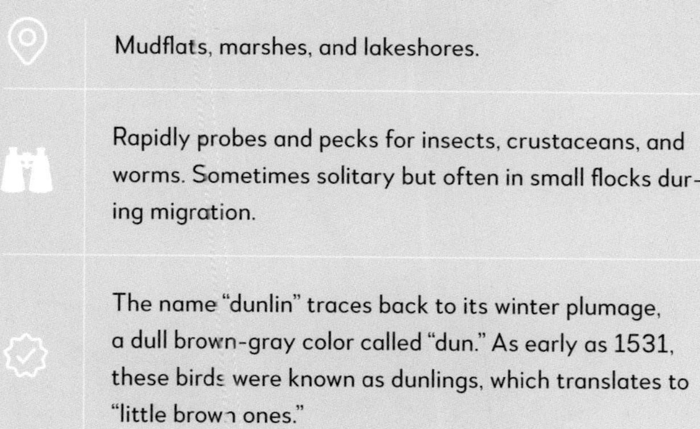

Mudflats, marshes, and lakeshores.

Rapidly probes and pecks for insects, crustaceans, and worms. Sometimes solitary but often in small flocks during migration.

The name "dunlin" traces back to its winter plumage, a dull brown-gray color called "dun." As early as 1531, these birds were known as dunlings, which translates to "little brown ones."

LENGTH: 6–8" / WINGSPAN: 14–15"

Breeding.
Tiny. Slightly drooping bill. Yellow legs.

LEAST SANDPIPER

Calidris minutilla

Tiny shorebird with yellowish legs, gently curved bill. In breeding season, reddish-brown back contrasts with white underparts and delicately streaked breast.

Mudflats, marshes, and shallow ponds.

Highly gregarious, forming loose flocks during migration and while foraging. Probes mud for insects, crustaceans, and mollusks.

The Least Sandpiper is the smallest sandpiper species in the Great Lakes region, measuring only 5 to 6 inches long and weighing roughly 1 ounce.

LENGTH: 5–6" / WINGSPAN: 10–11"

Breeding.
Long sturdy bill. Buff and brown striped head.

SHORT-BILLED DOWITCHER

Limnodromus griseus

Medium-sized shorebird with chunky body, long bill, and mottled brown feathers accentuated with black and red-orange tones. Underparts white with orange breast during breeding season.

Mudflats, marshes, small lakes, and shallow ponds.

Probes mudflats with a sewing machine head motion while foraging for small invertebrates, insects, crustaceans, worms, and mollusks.

The Short-billed Dowitcher has nerves at the tip of its beak that aid in locating food.

LENGTH: 9–11" / WINGSPAN: 19"

Adult.
Large eye. Cryptic brown plumage. Short legs.

AMERICAN WOODCOCK

Scolopax minor

Enigmatic and cryptically colored shorebird with compact body; round head; long, straight bill; and large eyes set far back. Also known as the timberdoodle.

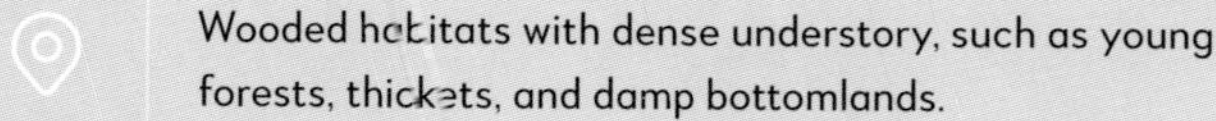

Wooded habitats with dense understory, such as young forests, thickets, and damp bottomlands.

Rocks back and forth and probes moist soil with sensitive bill, foraging for earthworms and other invertebrates. During breeding season, gives unique *peent* call while on ground, then takes flight for aerial display.

Males showcase their prowess with "sky dances," spiraling flights accompanied by chirps and twitters over open fields or forest clearings. Their mesmerizing descent in a fluttering display includes melodic sounds, an aerial performance sure to impress potential mates.

LENGTH: 9–12" / WINGSPAN: 16–18"

Adult.
Long bill. Boldly patterned face.

WILSON'S SNIPE

Gallinago delicata

Medium-sized chunky shorebird with boldly striped head and back, awkwardly long bill, white belly with dark barring on sides, and mottled brown-and-buff plumage.

Wetlands, flooded fields, marshes, pond and stream edges. Prefers dense vegetation such as cattails, sedges, and reeds.

Uses long bill to probe mud for invertebrates. Shy and well-camouflaged, hides in vegetation. Flushes if approached, employing aerial zigzags to evade predators. Mostly solitary but can be found in loose flocks.

During a steep dive in the snipe's courtship flight display, air rushes through its fanned tail feathers and produces a high-pitched, wavering sound called winnowing.

LENGTH: 10–13" / WINGSPAN: 16–17"

Breeding.
Spotted chest, orange bill with black tip. Often bobs tail.

SPOTTED SANDPIPER

Actitis macularius

Small brown sandpiper with dark spots on white belly. White line over eyes, orange bill with black tip. Usually solitary.

Along rivers, streams, ponds, and lakeshores.

Feeds on insects, crustaceans, and aquatic larvae. Practices polyandry where females mate with multiple males during a breeding season. Females choose territories, perform courtship displays, and initiate copulation.

Spotted Sandpipers exhibit a distinctive teetering motion as they walk along the shore.

LENGTH: 7–8" / WINGSPAN: 14–16"

Breeding.
Mottled body, white belly, long yellow legs. Thin bill.

LESSER YELLOWLEGS

Tringa flavipes

Medium-sized shorebird with mottled white, gray, and brown plumage; a thin, needlelike bill; and bright yellow legs characteristic of the species.

Mudflats, marshes, shorelines, and shallow ponds during migration.

Probes mud for small insects, crustaceans, and other invertebrates. Walks quickly or runs through shallow water, pecking and jabbing for small fish. Give a high-pitched *too-too* call in flight.

The Lesser Yellowlegs bobs its head up and down as it watches an intruder, demonstrating its alertness.

LENGTH: 9–10" / WINGSPAN: 23–25"

Nonbreeding.
Light gray all over. Straight black bill.

WILLET

Tringa semipalmata

Medium-sized shorebird with sleek gray plumage, black-and-white wings in flight, pale gray legs, and a long, straight bill. Long-distance migrant between breeding and wintering grounds.

Sandy beaches, mudflats, and marshes.

Uses long bill to probe substrate for worms, crustaceans, and mollusks.

It is named for its loud call, *pill-will-willet*.

LENGTH: 13–16" / WINGSPAN: 27"

Adult.
White body, black wingtips.
Yellow bill with black ring.

RING-BILLED GULL

Larus delawarensis

Medium-sized gull with white body, gray wings, black wing-tips, yellow legs and feet. Bill yellow with black ring near the tip. Dark ring around eye during breeding season. Juveniles speckled brown and gradually change to adult plumage as they mature.

Lakes, rivers, beaches, parking lots, and landfills.

Highly social, highly vocal, adaptable omnivore, especially successful in urban environments. Feeds on fish, insects, small mammals, and crustaceans, and scavenges at garbage dumps and picnic areas.

The black ring near the tip of the Ring-billed Gull's yellow bill becomes more prominent during the breeding season, making the bird easily recognizable among gull species in its range.

LENGTH: 16–21" / WINGSPAN: 41–46"

Adult.
Snowy white with pale gray back. Yellow bill with red spot.

Adult.
Black wingtips.

HERRING GULL

Larus argentatus

Large gull with white body, gray wings and back, black wing-tips, pink legs and feet. Bill yellow with red spot near the tip.

Beaches, lakes, mudflats, harbors, picnic areas, parking lots and garbage dumps.

Highly social, highly vocal, opportunistic omnivore. Feeds on fish, crustaceans, mollusks insects, small mammals, and carrion as well as scavenging in urban areas, land-fills, and picnic grounds.

Herring Gulls are intelligent birds with complex social behaviors. They demonstrate problem-solving skills by dropping hard-shelled prey from a height to crack it open.

LENGTH: 22–26" / WINGSPAN: 53–57"

Adult.
Slim bill. Black head and chest; gray back, wings, and tail.

BLACK TERN

Chlidonias niger

Small tern with predominantly black breeding plumage, contrasting with gray wings and white underbelly.

Shallow marshes, ponds, lakeshores, and wetlands with emergent vegetation.

Feeds on small fish, aquatic invertebrates, and insects. Impressive and agile aerialist whether hunting or courting. Builds a floating nest in shallow water within a breeding colony.

Black Terns give sharp, high-pitched calls and trills. These vocalizations communicate between mates, signal territorial defense, and warn other terns.

LENGTH: 9–14" / WINGSPAN: 22–23"

Male.
Colorful bare skin.

Male.
Large dark body and tiny head.
Bare upper neck and head.

WILD TURKEY

Meleagris gallopavo

Very large game bird with iridescent feathers in shades of bronze, copper, and green on male, dark brown with darker barring on female. Bare, fleshy, blue-and-red head and neck can change color depending on mood or social status.

Open forests, woodlands, grasslands, and agricultural areas.

Feeds on seeds, fruits, nuts, insects, and small vertebrates. During breeding season, males puff up, fan the tail feathers, and produce distinctive gobbling calls to attract mates.

Despite being quite large, turkeys are surprisingly adept at flying and commonly roost in trees.

LENGTH: 43–45" / WINGSPAN: 49–56"

Female.
Small head, large body.
Cryptic coloration.

Male.
Orange eyebrows and air sacs.
Erect head feathers.

GREATER PRAIRIE-CHICKEN

Tympanuchus cupido

Medium-sized grouse with dark brown barred plumage, small head, short tail, round body. Male has orange eyebrows and orange air sacs on neck during courtship displays.

Historically associated with tallgrass prairies and native grasslands; now found in remnants of native prairie, grassy meadows, and agricultural fields with suitable habitat.

During breeding season, males gather at a communal display area (lek) to perform. They inflate orange air sacs, erect head feathers, and produce deep, booming calls to establish dominance and compete for mating opportunities. Females select males based on display and vocalization skills, and then choose a suitable nesting site.

Greater Prairie-Chickens are nonmigratory, residing within their breeding range year-round. They are an indicator species for grassland conservation, highlighting the need to preserve and restore native grassland habitats for survival.

LENGTH: 16" / WINGSPAN: 27–28"

Male.
Bronzy plumage. Long tail.
Red skin around eyes.

RING-NECKED PHEASANT

Phasianus colchicus

Ground-dwelling game bird with iridescent shades of green, red, and bronze in male; female cryptically colored in pale buff brown. Both have long tail feathers. Named for white ring around male's neck.

Agricultural fields, grasslands, marshes, wetlands, and brushy areas with suitable cover.

Feeds on seeds, grains, insects, and small invertebrates. Secretive, often running rather than flying when disturbed. During breeding season, males strut with erected tail feathers and call loudly to attract females.

The Ring-necked Pheasant is a master of ground living, often taking to the air only as a last resort when no other escape route is available.

LENGTH: 19–27" / WINGSPAN: 22–33"

Juvenile.
Featherless black head. Black body.

Adult.
Featherless red head. Gray flight feathers.

TURKEY VULTURE

Cathartes aura

Large black body contrasts sharply with featherless red head and pale, hooked beak. Pale gray wing feathers visible in flight.

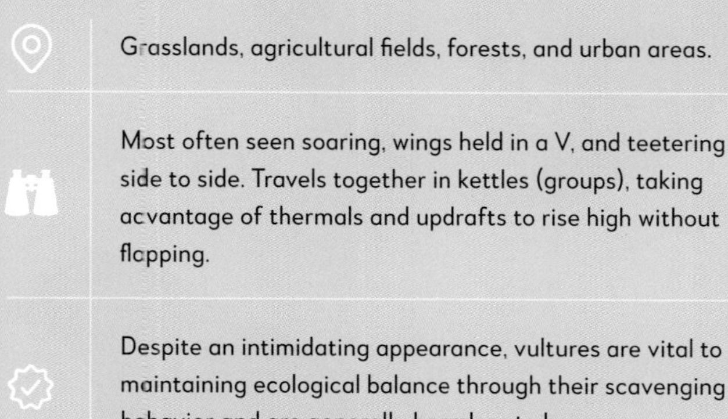

Grasslands, agricultural fields, forests, and urban areas.

Most often seen soaring, wings held in a V, and teetering side to side. Travels together in kettles (groups), taking advantage of thermals and updrafts to rise high without flapping.

Despite an intimidating appearance, vultures are vital to maintaining ecological balance through their scavenging behavior and are generally harmless to humans.

LENGTH: 25–31" / WINGSPAN: 66–70"

Adult.

Dark back and wings, white head and throat, dark stripe through yellow eye. Strongly hooked bill.

Adult.

Mostly white below with dark patches at bend in wing (wrist).

OSPREY

Pandion haliaetus

Predominantly white head, chest, and underparts. Dark brown wings, back, and eyeline. In flight, wings bent at the wrist, resembling a flying M or W, depending on angle.

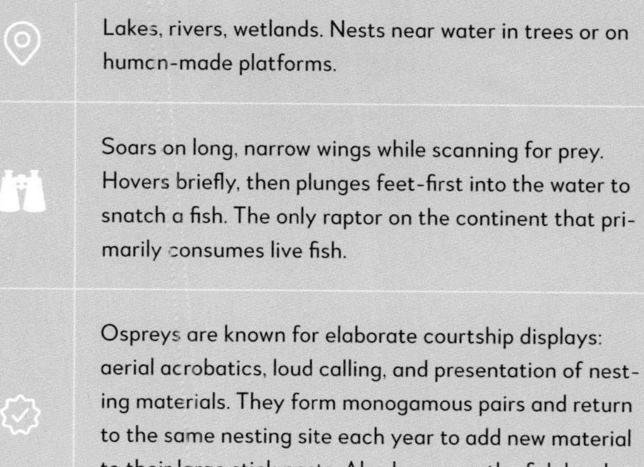

Lakes, rivers, wetlands. Nests near water in trees or on human-made platforms.

Soars on long, narrow wings while scanning for prey. Hovers briefly, then plunges feet-first into the water to snatch a fish. The only raptor on the continent that primarily consumes live fish.

Ospreys are known for elaborate courtship displays: aerial acrobatics, loud calling, and presentation of nesting materials. They form monogamous pairs and return to the same nesting site each year to add new material to their large stick nests. Also known as the fish hawk.

LENGTH: 21–22" / WINGSPAN: 59–70"

Adult male.
Owl-like face. Gray head, speckled underparts. Slender wings.

NORTHERN HARRIER

Circus hudsonius

Slender body, long wings, long banded tail decorated with a white rump. Owl-like facial appearance. Notable sexual dimorphism: adult males grayish with black wingtips; adult females rich brown.

Grasslands, marshes, meadows, wetlands, and agricultural fields.

Feeds on voles, mice, and rabbits, but also takes birds, reptiles, and insects. Flies low and slow over open areas, using keen eyesight and hearing to locate and capture prey with sharp talons.

The male is nicknamed "gray ghost" because of its striking bluish-gray upper body and contrasting white underparts and is less frequently seen than the female, adding a touch of mystery.

LENGTH: 18–19" / WINGSPAN: 40–46"

Adult.
Sexes similar but female is larger.
Dark cap, red-and-white breast.
Long, rounded, banded tail.

COOPER'S HAWK

Accipiter cooperii

Medium-sized raptor with slate-gray back and wings, rusty-orange barred chest and belly (adult), and a long, rounded tail with narrow white bands. First-year birds have brown streaks on light chest, brown back with white spots.

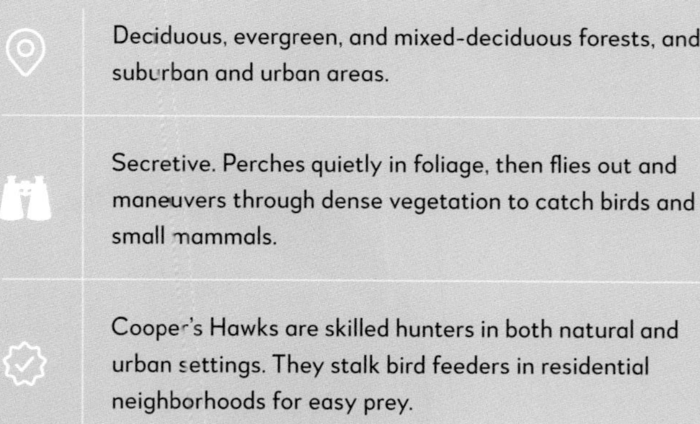

Deciduous, evergreen, and mixed-deciduous forests, and suburban and urban areas.

Secretive. Perches quietly in foliage, then flies out and maneuvers through dense vegetation to catch birds and small mammals.

Cooper's Hawks are skilled hunters in both natural and urban settings. They stalk bird feeders in residential neighborhoods for easy prey.

LENGTH: 14–16" / WINGSPAN: 24–35"

Adult.
Sexes similar but female is larger. Huge wingspan. White head and tail. Obvious primary feather "fingers."

Adult.
Massive yellow bill. White head.

BALD EAGLE

Haliaeetus leucocephalus

Unmistakable in adult plumage. White head and tail contrast with dark brown body. Massive yellow bill. Wingspan up to 7 feet. Birds less than four or five years old are varying degrees of brown and splotchy white.

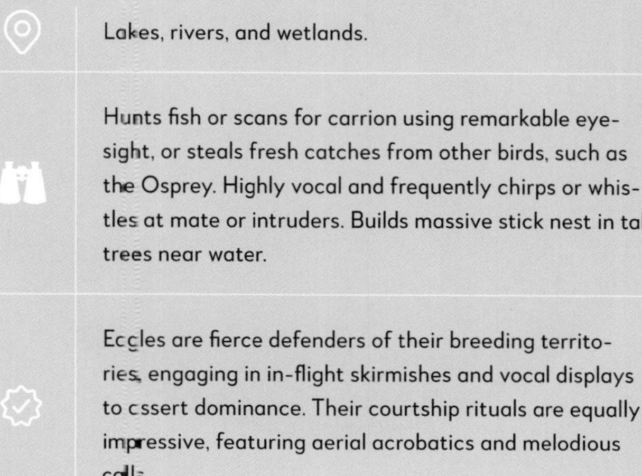

Lakes, rivers, and wetlands.

Hunts fish or scans for carrion using remarkable eyesight, or steals fresh catches from other birds, such as the Osprey. Highly vocal and frequently chirps or whistles at mate or intruders. Builds massive stick nest in tall trees near water.

Eagles are fierce defenders of their breeding territories, engaging in in-flight skirmishes and vocal displays to assert dominance. Their courtship rituals are equally impressive, featuring aerial acrobatics and melodious calls.

LENGTH: 27–37" / WINGSPAN: 84"

Adult.
Red breast and shoulders. Striking black-and-white wings.

RED-SHOULDERED HAWK

Buteo lineatus

Medium-sized reddish-brown raptor with reddish barred chest and belly. Broad black-and-white checkered wings and banded tail stand out in flight.

Mixed woodlands near streams and rivers. Prefers mature trees for nesting and open spaces for hunting.

Preys on mice, voles, rabbits, amphibians, reptiles, and occasionally small birds. Hunts by perching in elevated locations, scanning for prey, and swooping to capture it with sharp talons.

Red-shouldered Hawks are not just visually striking; they also have a unique voice. Their loud, piercing calls, often likened to *kee-yer*, serve various purposes, such as communication between mates, territorial defense, and courtship displays.

LENGTH: 16–24" / WINGSPAN: 37–43"

Juvenile.
Dark head, light chest, dark bellyband. Light eye indicates juvenile. Adult has dark eye, red tail, and same pattern on chest.

RED-TAILED HAWK

Buteo jamaicensis

Bulky hawk with red tail feathers. Dark brown back and wings, lighter underbelly with darker band. In flight from below, dark patagial marks (leading edge of wing) are obvious. Birds less than three or four years old have barred tail but still show dark patagials and belly band.

Open fields, grasslands, agricultural areas, and forested regions with adjacent open spaces. Also suburban and urban areas with suitable prey.

Feeds on rodents, rabbits, and ground squirrels. Hunts by soaring high, then swooping down to catch prey with sharp talons. Gives loud, piercing screams, especially during flight or while perched on high vantage points.

The Red-tailed Hawk is the most widespread and frequently encountered hawk species in the United States, but its plumage is highly variable, with western and eastern birds looking quite different from each other.

LENGTH: 17–25" / WINGSPAN: 37–43"

Male.
Dark facial markings. Slate-blue wings and crown.

Female.
Brown plumage with dark bars. Light facial markings.

AMERICAN KESTREL

Falco sparverius

Small yet fierce bird of prey adorned in reddish-brown cloak with striking black markings on face, wings, and tail. Male's wings slate-blue; female's wings heavily barred.

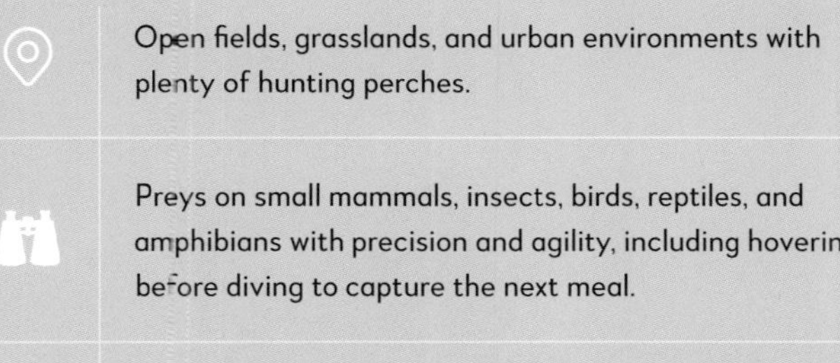

Open fields, grasslands, and urban environments with plenty of hunting perches.

Preys on small mammals, insects, birds, reptiles, and amphibians with precision and agility, including hovering before diving to capture the next meal.

Migration adds a dynamic dimension to the kestrel's behavior. While some individuals make short-distance journeys in response to changing prey availability or weather patterns, others are year-round residents, demonstrating adaptability and resilience to changing environments.

LENGTH: 8–12" / WINGSPAN: 20–24"

Adult.
Gray body, broad shoulders, iridescent nape.

ROCK PIGEON

Columba livia

Stout body, short legs, small, rounded head with small, dark bill. Plumage varies widely, from gray and white to more colorful variations with iridescent feathers on neck and wings.

Urban and suburban areas, also agricultural fields, grasslands, and cliffs. Highly adaptable and take advantage of human-altered landscapes.

Feeds on seeds, grains, and plant matter foraged on the ground or gleaned from agricultural fields and urban environments. Supplements diet with small insects and invertebrates. Known for cooing calls and habit of roosting and nesting on ledges, buildings, bridges, and other structures.

Rock Pigeons have been domesticated for thousands of years and were bred for racing, for exhibition, and as pets. Their homing talents have been utilized by humans for messaging and communication.

LENGTH: 11–14" / WINGSPAN: 19–26"

Adult.
Small head on large body. Long, pointed tail. Gray-brown plumage, black spots on wings.

MOURNING DOVE

Zenaida macroura

Small head, light blue eyering, thin black bill, rounded body, and long, tapered tail. Soft gray-brown plumage with lighter peach underparts and scattered black spots on wings.

Open woodlands, fields, and grasslands, urban and suburban areas. Highly adaptable but prefers scattered trees or shrubs. May even nest in a hanging flower basket.

Feeds on seeds, grains, and small fruits. Often forages on ground, or perches while surveying for potential food sources. Pairs exhibit gentle and peaceful behavior, sitting together and cooing softly.

Mourning Doves have an impressive reproductive capacity, often raising up to six broods per year.

LENGTH: 9–13" / WINGSPAN: 17–18"

Adult gray morph.
Small ear tufts not always visible. Gray face bordered by black. Gray-and-black cryptic plumage.

Adult red morph.
Similar cryptic plumage but with reddish tones.

EASTERN SCREECH-OWL

Megascops asio

Small owl with mottled gray, reddish-brown, or brown plumage that blends seamlessly with tree bark. Ear tufts, rounded head, and bright yellow eyes.

Forests, woodlands, parks, and suburban areas with mature trees.

Primarily nocturnal, using keen hearing and sharp talons to locate and capture small mammals, birds, and insects, and frogs, lizards, and fish when available.

Eastern Screech-Owls do not screech. Their vocalizations include soft, descending whistles known as whinnies, as well as trills and barks used for communication and territorial defense.

LENGTH: 6–9" / WINGSPAN: 18–24"

Adult.
Catlike face with yellow eyes. Ear tufts.

GREAT HORNED OWL

Bubo virginianus

Large and formidable bird of prey with "horns" (actually feather tufts) and piercing yellow eyes. Dense, mottled brown plumage provides excellent camouflage in woodland habitats.

Forests, woodlands, open fields, marshes, and urban areas with suitable roosting and nesting sites. Prefers mature trees and dense vegetation.

Feeds on mammals, birds, reptiles, amphibians, and large insects. Silent flight, keen hearing, and powerful talons allow ambush and capture of prey with precision. Predominantly nocturnal, hunting at night and resting during the day.

The Great Horned Owl is named for the tufts on its head that resemble horns, giving it a majestic and imposing presence in the natural world.

LENGTH: 18–24" / WINGSPAN: 39–57"

Male.
Snow-white plumage, bright yellow eyes. Amount of dark barring depends on age and sex.

SNOWY OWL

Bubo scandiacus

Large owl with predominantly white plumage, piercing yellow eyes, and large, powerful talons. Variable dark barring and spots on otherwise white body.

In winter, open fields near large bodies of water, tundra-like habitats, agricultural fields, and sometimes fields near airports.

Diurnal. Often perches on ground, low vegetation, or buildings. Feeds on rodents and small birds. Uses keen eyesight to spot movement from a distance and powerful talons to secure the catch.

The Snowy Owl is the heaviest owl species in North America. It needs large body mass and substantial feather insulation to survive Arctic temperatures.

LENGTH: 20–27" / WINGSPAN: 49–57"

Juvenile.
Fluffy round head. Yellow bill. Dark eyes.

Adult.
Round head. Dark eyes. Yellow bill. Horizontal dark bars on chest, vertical dark streaks on lower body.

BARRED OWL

Strix varia

Large owl with round head, dark eyes, pale facial disk bordered by dark bars, and prominent vertical streaks on chest.

Mature deciduous and mixed forests with dense vegetation and ample tree cover. Also urban and suburban wooded areas.

Primarily nocturnal, using keen hearing and silent flight to locate and capture small mammals, birds, amphibians, and invertebrates. Day-roosting owls often mobbed by Blue Jays and other songbirds.

Barred Owls emit a haunting call that sounds like *who-cooks-for-you, who-cooks-for-you-all.* These calls serve multiple purposes, including territorial defense, communication between mates, and advertising their presence to potential mates.

LENGTH: 16–19" / WINGSPAN: 39–43"

Adult.
Big round head and short tail present a front-heavy look in flight.

SHORT-EARED OWL

Asio flammeus

Round head with indistinct ear tufts, white face, and unique facial disk featuring black eye patches around bright yellow eyes. Overall plumage is mottled and cryptic, providing excellent camouflage.

Open grasslands, marshes, and agricultural fields. Prefers low vegetation.

Hunts at dawn, dusk, or night, flying low over open fields and scanning for mice and meadow voles, among other small mammals.

The Short-eared Owl is one of the few owl species that nests on the ground.

LENGTH: 13–16" / WINGSPAN: 33–40"

Adult.
White V between eyes. White spots on wings. Rust-colored stripes on belly.

Adult.
Round head. Yellow eyes. Dark bill.

NORTHERN SAW-WHET OWL

Aegolius acadicus

Small nocturnal owl with round head, large yellow eyes, and brownish-gray plumage. Feathers marked with white streaks, giving mottled appearance to camouflage it against tree bark.

Coniferous and mixed woodlands and wooded areas near rivers, streams, and marshes.

Feeds on mice, voles, and shrews along with insects and other invertebrates. Adept hunter, using silent flight and keen senses to locate and capture prey in darkness. Call resembles the sound of a saw being sharpened.

Northern Saw-whet Owls have remarkable endurance and navigation skills, as they can cross vast bodies of water such as the Great Lakes during their seasonal migrations.

LENGTH: 7–8" / WINGSPAN: 16–18"

Female.
Ragged crest. Big head. Daggerlike bill. Blue-gray with white neck and belly. Female has rusty red belt and sides.

BELTED KINGFISHER

Megaceryle alcyon

Compact body, large head, daggerlike bill, and shaggy crest. Blue-gray back and wings, white breast and belly, blue-gray band across chest. Female has additional rust-colored belt below chest band.

Rivers, lakes, and streams. Always near water. Needs steep, earthen banks or cliffs to excavate nesting burrows and roosting sites.

Highly specialized fish hunter. Perches over water, then dives headfirst with remarkable speed and precision to capture fish, crustaceans, and aquatic insects. Gives rattling call during territorial disputes or in warning.

During the breeding season, Belted Kingfishers court each other with aerial acrobatics, mutual preening, and fish gifts between mates.

LENGTH: 11–13" / WINGSPAN: 18–22"

Adult.
Brilliant red head. White body. Black wings with white secondary feathers. Note zygodactyl feet typical of woodpeckers.

RED-HEADED WOODPECKER

Melanerpes erythrocephalus

Striking glossy black back contrasts sharply with vibrant red head and neck. Prominent white wing patches obvious in flight.

Deciduous woodlands, forest edges, orchards, parks, and even suburban areas. Needs an abundance of dead wood and prefers open spaces in which to efficiently hunt flying insects.

Feeds on insects, fruits, nuts, seeds, and occasionally small vertebrates. Powerful flier. Strength and aerial prowess match aggressive nature. Uses robust bill to excavate holes in trees to find insects and larvae.

Red-headed Woodpeckers store food in crevices or holes in trees, a behavior known as caching. They may store hundreds of insects, seeds, and nuts in a single cache, which they rely on during times of scarcity or when caring for their young.

LENGTH: 7–9" / WINGSPAN: 16"

Male.

Tan body. Black-and-white wings and striped back. Male's red crown reaches bill; female's is shorter.

RED-BELLIED WOODPECKER

Melanerpes carolinus

Striking black-and-white barred back topped with a vibrant red crown. Red belly usually subtle, often just a hint on lower belly.

Deciduous forests, wooded suburban areas, and parks with mature trees.

Feeds on insects, fruits, nuts, seeds, and sometimes small vertebrates. Uses strong bill to hammer into tree bark for food and excavate nest cavities. Typical *churr* call and rhythmic drumming used for communication, territory defense, and courtship displays.

The Red-bellied Woodpecker's tongue is barbed and its saliva is sticky, helping it efficiently snatch prey from deep crevices in tree bark.

LENGTH: 9" / WINGSPAN: 13–16"

Male.
Red throat. White stripe over bill and face.

Juvenile.
Bold white stripe in wing diagnostic for all ages and sexes.

YELLOW-BELLIED SAPSUCKER

Sphyrapicus varius

Black-and-white barred upperparts, red crown and throat patch in males (white throat patch in females), and bright yellow wash on underparts.

Mixed forests, woodlands, and forest edges with deciduous and coniferous trees. Attracted to sap-producing trees such as maple, birch, and fruit trees.

Highly specialized feeding behavior: drills rows of shallow holes called sapwells; consumes oozing sap and insects attracted to sap. Also takes fruits and berries. Known for irregular drumming rhythm and nasal mewing calls during courtship and territorial displays.

Sapwells not only provide the sapsucker with food but benefit other birds, such as hummingbirds, that feed on both the sap and the insects attracted to the sap.

LENGTH: 7–8" / WINGSPAN: 13–15"

Male.
White belly. Checkered wings. Dainty, chisel-like bill. Red patch on head indicates a male.

DOWNY WOODPECKER

Dryobates pubescens

Small woodpecker with black-and-white plumage. White belly, black wings with white spots, and white back with black bars. Most prominent feature is short, chisel-like bill.

Deciduous and mixed forests, woodlands, parks, and suburban areas with mature trees. Prefers mix of open spaces and dense vegetation.

Uses strong bill to peck and drill into tree bark to find insects. Skilled climber, hopping and spiraling up trunks and branches in search of food.

Downy Woodpeckers give a distinctive whinny call to communicate between mates and family members and for territorial displays. They also drum on trees to communicate with other woodpeckers.

LENGTH: 5–6" / WINGSPAN: 9–11"

Male.
Checkered wings. White belly. Sturdy bill. Red patch on head indicates a male.

HAIRY WOODPECKER

Dryobates villosus

Medium-sized woodpecker with black-and-white plumage. Bold black markings on back and wings, relatively large chisel-like bill, white belly, and white stripe extending from base of bill to neck.

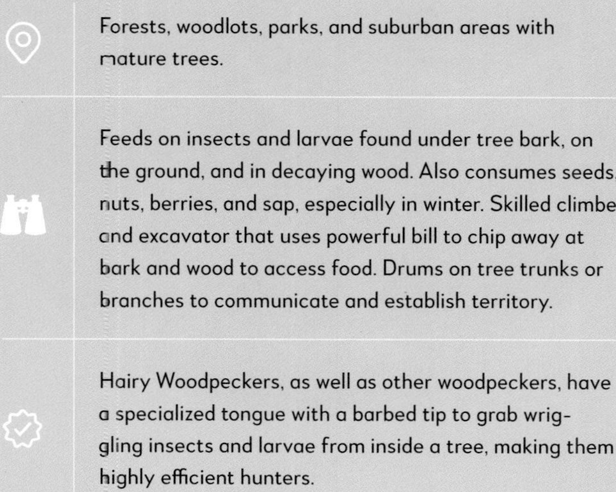

Forests, woodlots, parks, and suburban areas with mature trees.

Feeds on insects and larvae found under tree bark, on the ground, and in decaying wood. Also consumes seeds, nuts, berries, and sap, especially in winter. Skilled climber and excavator that uses powerful bill to chip away at bark and wood to access food. Drums on tree trunks or branches to communicate and establish territory.

Hairy Woodpeckers, as well as other woodpeckers, have a specialized tongue with a barbed tip to grab wriggling insects and larvae from inside a tree, making them highly efficient hunters.

LENGTH: 7–10" / WINGSPAN: 13–16"

Female.
Brown with black barring. Gray nape with red patch. Male similar but with black mustache.

NORTHERN FLICKER

Colaptes auratus

Brownish woodpecker with black bib on chest, red patch on nape, and black spots on underside. Wing and tail feathers have bright yellow shafts, visible in flight. Male has stylish black mustache.

Open woodlands, forest edges, parks, and suburban areas.

Excels at ground foraging, primarily for insects, especially ants. Engages in courtship displays like head-weaving and body-bobbing, while aggressive head movements signal territorial defense. Very vocal, giving rhythmic *wik-a wik-a wik-a* or single *klee-yah*. Calls plus resonant drumming attracts mates and asserts presence.

The Northern Flicker is a fascinating example of how species can vary across regions. The western Red-shafted and eastern Yellow-shafted flickers were once considered separate species but were eventually "lumped" together because they interbreed where their ranges overlap.

LENGTH: 11–12" / WINGSPAN: 16–20"

Male.
Brilliant red crest atop black-and-white head and black body. Less vibrant red mustache. Female with black forehead, mustache.

PILEATED WOODPECKER

Dryocopus pileatus

Extra-large black woodpecker with bold white stripes on face and neck. Brilliant red crest gives regal look.

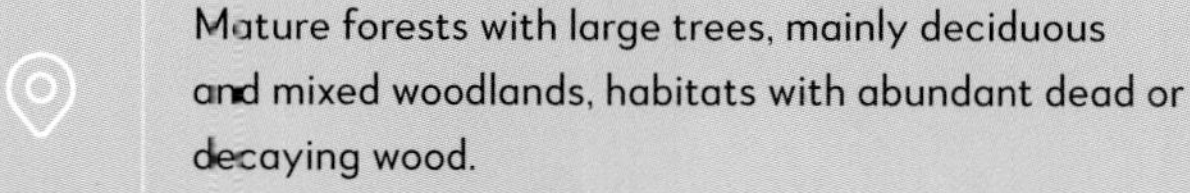

Mature forests with large trees, mainly deciduous and mixed woodlands, habitats with abundant dead or decaying wood.

Feeds on wood-boring insects, ants, and other invertebrates found within decaying wood. Uses strong bill to excavate deep into trees, then deploys barbed tongue to grab hidden prey. Loud, resonant calls and powerful drumming echo through forest and announce presence to rivals or mates.

The Pileated Woodpecker is the largest remaining woodpecker species in North America. The Ivory-billed Woodpecker (*Campephilus principalis*), now extinct, was about 19–20 inches long.

LENGTH: 15–19" / WINGSPAN: 26–29"

Adult.
Crested, with black necklace. Blue-and-white wings and tail. Blue-gray body.

BLUE JAY

Cyanocitta cristata

Vibrant blue plumage, black necklace, and bold black-and-white markings on wings and tail. Crested head.

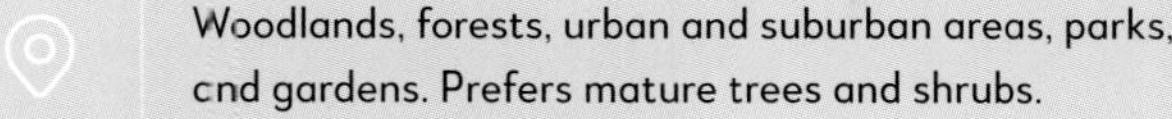

Woodlands, forests, urban and suburban areas, parks, and gardens. Prefers mature trees and shrubs.

Adaptable, assertive, and opportunistic. Takes insects, seeds, nuts, fruits, small vertebrates, and will visit bird feeders for peanuts. Loudly vocalizes and aggressively interacts with other birds when defending territory or competing for food.

Blue Jays look blue, but their feathers are actually brown because they contain melanin, a dark pigment. The blue color is an optical effect caused by light scattering within the feathers, making them a fascinating example of nature's color illusions.

LENGTH: 9–11" / WINGSPAN: 13–16"

Adult.
All over black plumage. Sturdy bill.

AMERICAN CROW

Corvus brachyrhynchos

Large bird with glossy black plumage, stout bill, and cawing calls. Larger Common Raven (*Corvus corax*) has massive bill, fluffy throat feathers, diamond-shaped tail in flight, and is less social.

Urban areas, farmlands, and forests. Particularly successful near human habitation.

Displays remarkable adaptability by exploiting garbage bins, parking lots, and agricultural fields for food. Highly social, forming tight-knit communities for cooperative foraging, territorial defense, and communal roosting.

American Crows possess remarkable intelligence, demonstrated by their ability to recognize individual humans and use tools. Their keen observational skills and cognitive prowess make them fascinating birds to study.

LENGTH: 15–20" / WINGSPAN: 33–39"

Male.
Ruby-red throat.

Female.
Metallic green back. Gray throat and belly.

RUBY-THROATED HUMMINGBIRD

Archilochus colubris

Tiny hummingbird with iridescent green upperparts and white underside. Vibrant ruby-red throat distinguishes male; female has plain gray throat. Long, slender bill adapted for sipping nectar from flowers.

Open woodlands, gardens, parks, meadows, and neighborhoods with ample nectar-producing flowers and hummingbird feeders.

Primary food is flower nectar, but also consumes small insects and spiders. Highly territorial and aggressively defends feeding areas with aerial displays and chases to assert dominance. Wings make humming sound while hovering midair or darting from flower to flower.

With over 360 species, the Trochilidae is Earth's second most diverse bird family! In the Great Lakes region, the Ruby-throated Hummingbird holds the spotlight as the only native hummingbird species, bringing a flash of iridescent color to our eastern landscapes.

LENGTH: 2–3" / WINGSPAN: 3–4"

Adult.
Brownish-gray upperparts. White underparts with dark band.

Adult.
Tiny bill. Short tail.

BANK SWALLOW

Riparia riparia

Nimble swallow with slender body, small bill, and short, forked tail. Upper body brownish gray, throat and underparts white, chest with brown band.

Rivers, lakes, and ponds. Sandy or gravelly banks for colonial nesting.

Highly social, forms large groups during breeding season. Skilled flier, executing swift and agile maneuvers to catch flies, winged ants, and beetles midair.

Bank Swallows use their feet and bill to excavate nesting tunnels several feet into a bank to protect eggs and chicks from predators and inclement weather.

LENGTH: 4–5" / WINGSPAN: 9–13"

Adult.
Blue above, white below. Tiny bill.

Adult.
Long, slender wings.

TREE SWALLOW

Tachycineta bicolor

Head and back iridescent blue and green, with tiny bill, black mask, and bright white throat and belly. Males and females similar but females more subdued. During breeding season, male's colors become brighter and more vibrant.

Open habitat near marshes, meadows, lakeshores, and riversides.

Highly social and gregarious, often in large flocks during breeding season. Remarkable aerialists, darting and swooping to catch flying insects such as flies, beetles, and mosquitoes.

The Tree Swallow was named for its preference for nesting in tree cavities, but it also accepts artificial nest boxes.

LENGTH: 4–5" / WINGSPAN: 11–13"

Adult.
Grayish brown head and back, lighter underparts. No dark chest band as in Bank Swallow.

NORTHERN ROUGH-WINGED SWALLOW

Stelgidopteryx serripennis

Small swallow with plain brownish-gray plumage blending subtly into lighter underbelly. Short, square tail adds to compact appearance, distinguishing it from other swallow species. Minuscule hooks on leading edges of wings give it its name.

Lakeshores, riversides, marshes, and wetlands. Sandy banks or cliffs for nesting sites.

Feeds on flies, mosquitoes, and beetles, flying low over open water or perching on wires or branches near the water's edge to catch insects midair.

Instead of building mud nests, these swallows nest in burrows or crevices in sandy banks or cliffs, adding protection for their eggs and young.

LENGTH: 4–5" / WINGSPAN: 10–11"

Male.
Glossy purple-black all over.
Slightly peaked head.

PURPLE MARTIN

Progne subis

Large, streamlined body with hooked bill; long, tapered wings; and slightly forked tail. Male's plumage deep glossy purple. Females and immatures lighter gray and blue.

Open habitats near lakes, rivers, and marshes. Urban or suburban neighborhoods and parks.

Feeds on beetles, flies, mosquitoes, and dragonflies caught midair thanks to exceptionally agile aerial maneuvers. Very vocal, with chortling, chirping, bubbling calls given in flight or perched. Long-distance migrant that winters in South America and breeds into Canada.

North America's largest species of swallow nests in cavities and relies on human-made birdhouses, gourds, or specialized housing complexes.

LENGTH: 7" / WINGSPAN: 15–16"

Female and male.
Dark blue back, long tail. Underparts rusty red in male, lighter in female.

BARN SWALLOW

Hirundo rustica

Streamlined body; long, pointed wings; and deeply forked tail. Dark blue upperparts, orange underparts on male. Female underside lighter. Rust-colored forehead.

Meadows, marshes, agricultural lands, lakes, rivers, and ponds.

Highly social, gathers in large flocks during migration and breeding seasons. Annual migration spans thousands of miles between North America and Central/South America. Constructs cup-shaped nest from mud, grass, and other materials, attaching to vertical surfaces like buildings, bridges, or cliffs.

Barn Swallows are expert aerial hunters with wide mouths and specialized bristles around their bills that help them snatch prey from the air.

LENGTH: 5–7" / WINGSPAN: 11–12"

Adult.
Black cap and throat. White cheek. Gray-and-white body.

BLACK-CAPPED CHICKADEE

Poecile atricapillus

Small, active songbird with black cap and throat contrasting with white cheeks and belly. Gray back; wings and tail darker.

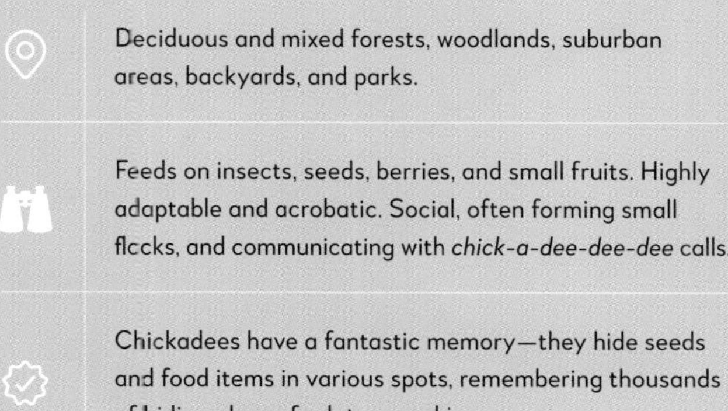

Deciduous and mixed forests, woodlands, suburban areas, backyards, and parks.

Feeds on insects, seeds, berries, and small fruits. Highly adaptable and acrobatic. Social, often forming small flocks, and communicating with *chick-a-dee-dee-dee* calls.

Chickadees have a fantastic memory—they hide seeds and food items in various spots, remembering thousands of hiding places for later snacking.

LENGTH: 4–5" / WINGSPAN: 6–8"

Adult.
Gray all over. Rusty wash on sides. Pointed crest.

TUFTED TITMOUSE

Baeolophus bicolor

Mostly gray bird with prominent black forehead, tufted gray crest, rusty peach-colored flanks, white underparts.

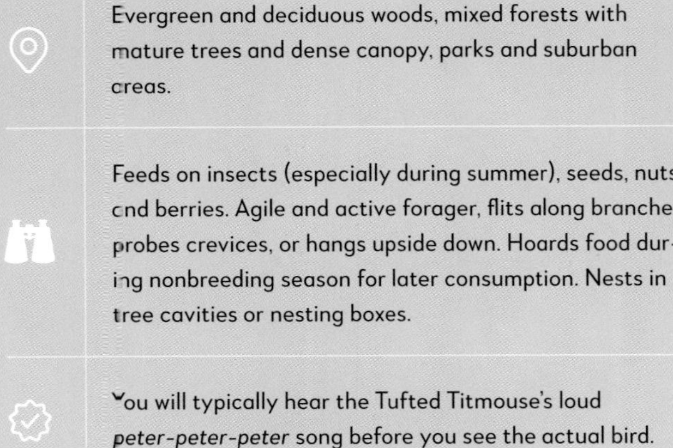

Evergreen and deciduous woods, mixed forests with mature trees and dense canopy, parks and suburban areas.

Feeds on insects (especially during summer), seeds, nuts, and berries. Agile and active forager, flits along branches, probes crevices, or hangs upside down. Hoards food during nonbreeding season for later consumption. Nests in tree cavities or nesting boxes.

You will typically hear the Tufted Titmouse's loud *peter-peter-peter* song before you see the actual bird.

LENGTH: 5–6" / WINGSPAN: 7–10"

Female.
Reddish sides. Blue-gray back. White eyebrow, black eyeline. Thin bill. Male has redder underparts.

RED-BREASTED NUTHATCH

Sitta canadensis

Small songbird with short tail, thin bill, blue-gray upperparts, and cinnamon-colored breast. White face and eyebrow with bold black stripe through eye.

Mature pine, spruce, and fir forests, mixed woodlands, suburban areas with suitable habitat.

Feeds on insects, spiders, and small invertebrates on tree bark and in crevices. Also takes seeds and nuts, using bill to wedge and crack them open. Nasal *yank yank yank* call and food-caching habit are notable behaviors.

The Red-breasted Nuthatch can climb down trees headfirst using its sturdy feet and claws, making it well-equipped to navigate its woodland habitat.

LENGTH: 4" / WINGSPAN: 7"

Adult.
White face and underparts. Slightly upturned slender bill. Blue-gray back. Short tail.

WHITE-BREASTED NUTHATCH

Sitta carolinensis

Compact body with short tail and slender bill. Slate-blue back, white face and underparts, black crown and nape.

Mature deciduous and mixed forests, woodlands, and suburban areas with ample tree cover.

Scoots headfirst down tree trunks and branches on large feet with sharp claws. Visits bird feeders for seed and suet. Forms monogamous pairs and excavates nest cavities in dead trees or abandoned woodpecker holes.

White-breasted Nuthatches do not migrate and are found in the Great Lakes region throughout the year.

LENGTH: 5" / WINGSPAN: 7–10"

Adult.
White belly; cryptically colored back and wings. White eyebrow. Thin, decurved bill.

BROWN CREEPER

Certhia americana

Small, well-camouflaged songbird with streaked brown-and-white plumage, white throat, and decurved bill.

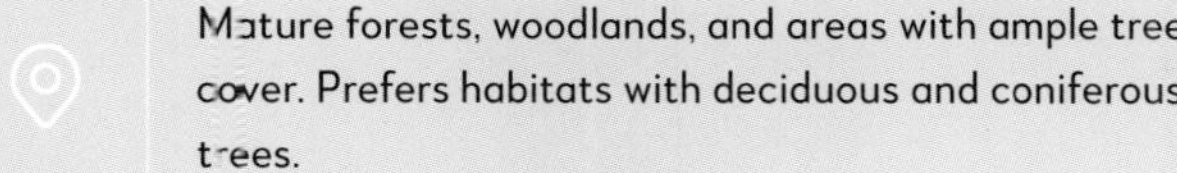

Mature forests, woodlands, and areas with ample tree cover. Prefers habitats with deciduous and coniferous trees.

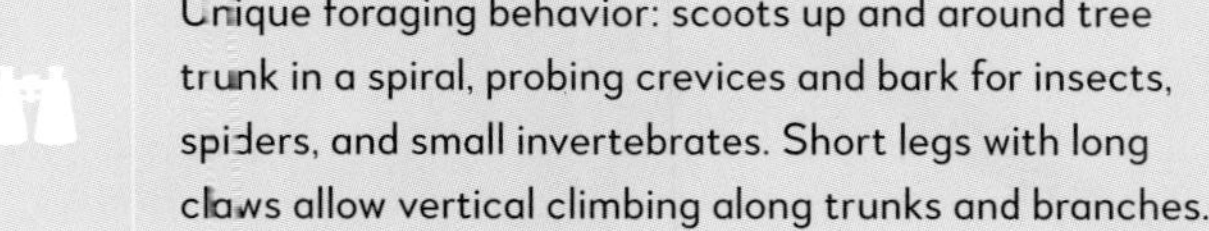

Unique foraging behavior: scoots up and around tree trunk in a spiral, probing crevices and bark for insects, spiders, and small invertebrates. Short legs with long claws allow vertical climbing along trunks and branches.

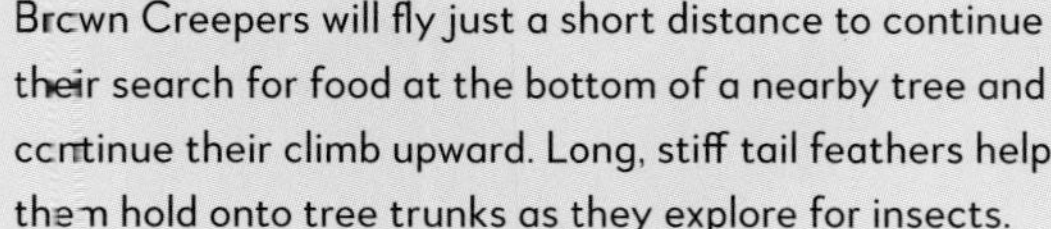

Brown Creepers will fly just a short distance to continue their search for food at the bottom of a nearby tree and continue their climb upward. Long, stiff tail feathers help them hold onto tree trunks as they explore for insects.

LENGTH: 4–5" / WINGSPAN: 6–7"

Adult.
Thin, pointed bill. White eyering. Long tail. Blue-gray upper plumage.

BLUE-GRAY GNATCATCHER

Polioptila caerulea

Small songbird with vibrant blue-gray upper plumage, white underbelly, long black tail edged with white, and prominent white eyering.

Deciduous and mixed woodlands, forest edges, parks, and gardens with dense vegetation. Prefers open canopy and dense understory vegetation.

Feeds on small insects, spiders, and other invertebrates found in foliage and branches. Highly active, agile forager. Catches insects midair with fast-paced, acrobatic movements, and hovers among leaves and branches while feeding.

Blue-gray Gnatcatchers are known for soft, whispery calls and high-pitched, nasal songs that consist of rapid, repetitive phrases. They use these vocalizations to establish territories, attract mates, and communicate with other gnatcatchers during the breeding season.

LENGTH: 4" / WINGSPAN: 6"

Adult.
Faint eyering and eyebrow. Grayish brown plumage with dark bars.

Adult.
Slightly decurved bill.

HOUSE WREN

Troglodytes aedon

Small, compact, agile songbird with plain brown plumage, pale eyebrow, lighter underparts, and slightly curved bill.

Woodlands, forests, parks, and suburban areas with suitable nesting cavities.

Feeds on insects and spiders but also berries and fruits, especially in fall and winter. Energetic and extremely territorial, flits and hops among branches while singing loudly. Skulks in dense vegetation while foraging.

Despite their small size, House Wrens are known for powerful and melodic songs, contributing to the lively chorus of bird sounds in woodlands and suburban areas.

LENGTH: 4–5" / WINGSPAN: 5"

Adult.
Streaked crown, light belly. Flexible.

Adult.
Streaked back, barred wings.

SEDGE WREN

Cistothorus stelllaris

Small, short-tailed songbird with streaked brown upperparts and buff-colored belly. Crown and wings streaked with white.

Open grasslands, meadows, wetlands, and fields with dense vegetation, mainly with abundant sedges and grasses. Nests near marshes, prairies, or agricultural areas.

Feeds on small insects, spiders, and other invertebrates. Acrobatic and agile. Hops and climbs within dense vegetation, probing grass stems and leaves. Rapid song consists of trills, buzzes, and chatters.

Employs a competitive nesting strategy by destroying the eggs of other birds in their territory.

LENGTH: 3–4" / WINGSPAN: 4–5"

Adult.

Light eyebrow, stripes on back. Black bars on tail. Often obscured by grass.

MARSH WREN

Cistothorus palustris

Small, secretive but fiery wren with short, barred tail and bold white eyebrow. Brown back streaked black and white. Pale underparts.

Marshes and wetlands with dense vegetation, such as cattails, reeds, and bulrushes. Prefers areas with tall emergent vegetation near the edges of marshes or wetlands.

Feeds on small insects and spiders. Exhibits remarkable aggression and vocal ability during breeding season, using complex songs to establish territories, attract mates, and communicate with other wrens.

Marsh Wrens are polygynous, with males forming relationships with multiple females in their territory and building multiple nests to accommodate them.

LENGTH: 3–5" / WINGSPAN: 5"

Adult.
Yellow feathers around eye give a spectacled look.

YELLOW-THROATED VIREO

Vireo flavifrons

Small, elegant songbird with distinctive yellow "spectacles," dark eye, and yellow throat. Two white wingbars on gray wings.

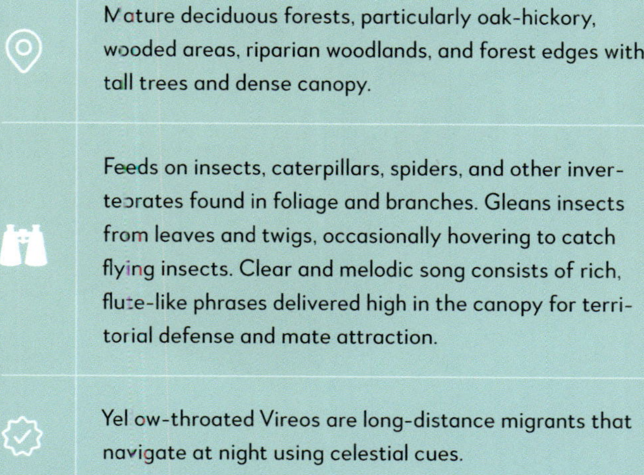

Mature deciduous forests, particularly oak-hickory, wooded areas, riparian woodlands, and forest edges with tall trees and dense canopy.

Feeds on insects, caterpillars, spiders, and other invertebrates found in foliage and branches. Gleans insects from leaves and twigs, occasionally hovering to catch flying insects. Clear and melodic song consists of rich, flute-like phrases delivered high in the canopy for territorial defense and mate attraction.

Yellow-throated Vireos are long-distance migrants that navigate at night using celestial cues.

LENGTH: 5" / WINGSPAN: 9"

Adult.
Grayish-olive plumage. Dark eye and eyeline, light eyebrow.

WARBLING VIREO

Vireo gilvus

Grayish-olive upperparts smoothly transition into whitish belly with hints of yellow on flanks. Prominent bill, dark eyes, light eyebrow, and faint eyeline.

Deciduous and mixed forests, woodlands, marshes along streams, and areas with dense foliage and mature trees. Also neighborhoods and parks with suitable habitat.

Feeds on insects hidden under leaves or branches high in treetops. May go unnoticed because of pale plumage, but song reveals its presence.

The Warbling Vireo is named after its fast, fluid, and repetitive warbling song.

LENGTH: 4–5" / WINGSPAN: 8"

Adult.
Red eye. Gray crown. Light eyebrow, black eyeline.

RED-EYED VIREO

Vireo olivaceus

Small olive-green vireo with whitish underparts, gray cap, white eyebrow, black eyeline. Red iris obvious.

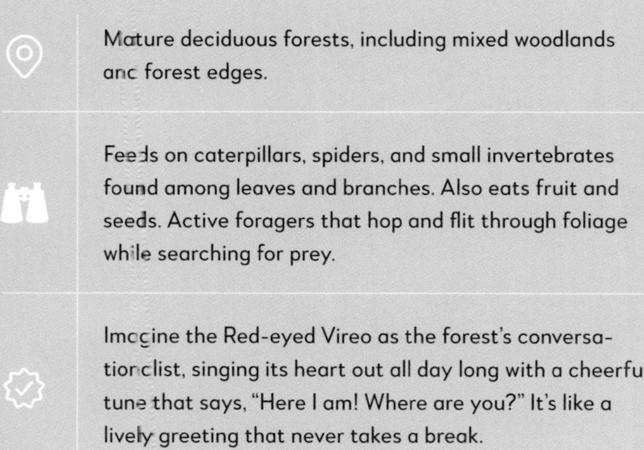

Mature deciduous forests, including mixed woodlands and forest edges.

Feeds on caterpillars, spiders, and small invertebrates found among leaves and branches. Also eats fruit and seeds. Active foragers that hop and flit through foliage while searching for prey.

Imagine the Red-eyed Vireo as the forest's conversationalist, singing its heart out all day long with a cheerful tune that says, "Here I am! Where are you?" It's like a lively greeting that never takes a break.

LENGTH: 4–5" / WINGSPAN: 9"

Male.
Red crown barely showing. Olive-greenish all over, tiny bill, white eyering and wingbar.

RUBY-CROWNED KINGLET

Corthylio calendula

Tiny songbird with tiny bill, olive-green upperparts, white eyering and wingbar, and subtle yellow accents on wings and tail.

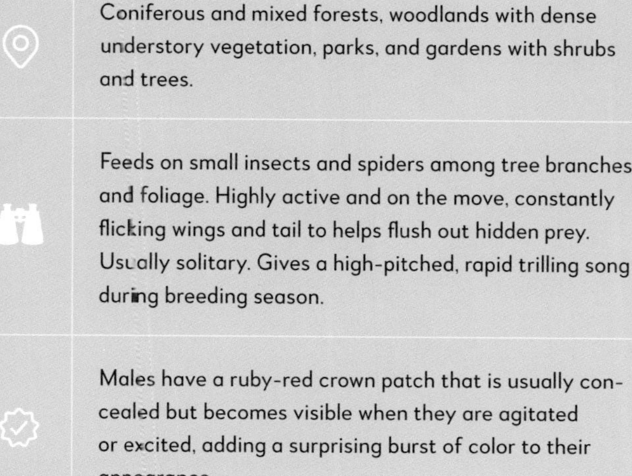

Coniferous and mixed forests, woodlands with dense understory vegetation, parks, and gardens with shrubs and trees.

Feeds on small insects and spiders among tree branches and foliage. Highly active and on the move, constantly flicking wings and tail to helps flush out hidden prey. Usually solitary. Gives a high-pitched, rapid trilling song during breeding season.

Males have a ruby-red crown patch that is usually concealed but becomes visible when they are agitated or excited, adding a surprising burst of color to their appearance.

LENGTH: 3–4" / WINGSPAN: 6–7"

Female.
Tiny and round. Yellow crown stripe bordered by black. White eyebrow. Male's crown stripe is orange.

GOLDEN-CROWNED KINGLET

Regulus satrapa

Tiny songbird with tiny bill, white eyebrow and wingbar, bright olive-green back, and pale gray underparts. Vibrant yellow-orange crown outlined by bold black stripes.

Coniferous forests, especially pine, spruce, and fir. Prefers thick vegetation.

Feeds on small insects and their larvae in tree foliage. Highly active, constantly moving and flitting. High-pitched, musical song consists of ascending, accelerating notes followed by descending warble.

Golden-crowned Kinglets are incredibly agile and can hover briefly while foraging for insects in the canopy, allowing them to access food sources out of reach for larger birds.

LENGTH: 3–4" / WINGSPAN: 5–7"

Adult.
Yellow belly. Olive-green upperparts. Rusty orange tail.

GREAT CRESTED FLYCATCHER

Myiarchus crinitus

Colorful and vocal flycatcher with crested head, vibrant lemon-yellow belly, olive-green back, and rusty orange tail feathers.

Deciduous forests, woodlands, parks, and suburban areas with mature trees. Prefers combination of open spaces for foraging and mature trees for nesting.

Feeds on flying insects such as beetles, moths, butterflies, and bees caught midair. Skilled aerial hunter. Gives loud and distinctive calls for territorial display, attracting mates, and communication.

Great Crested Flycatchers are resourceful architects, repurposing old woodpecker holes or natural cavities for their nesting sites.

LENGTH: 6–8" / WINGSPAN: 13"

Adult.
Slightly peaked, black head. Black face and back. White throat and belly.

EASTERN KINGBIRD

Tyrannus tyrannus

Medium-sized flycatcher with predominantly black plumage, white throat, breast, and belly. White outer tail feathers visible in flight.

Fields, meadows, grasslands, and wetlands. Prefers scattered trees and shrubs for perching and hunting insects.

Catches flying insects midair using sharp beak and agile flight. Also consumes berries and fruits when insect prey is scarce. Highly territorial during breeding season.

Eastern Kingbirds are not related to kings, and their name likely comes from their bold and aggressive behavior. Neither is their range strictly eastern, instead spanning almost all of North America.

LENGTH: 7–9" / WINGSPAN: 13–15"

Adult.
Flattened bill. Peaked head.
Yellowish belly. Wingbars.

EASTERN WOOD-PEWEE

Contopus virens

Small flycatcher with olive-gray plumage, lighter underbelly, and slightly crested head. Dark eyes and small, flattened bill.

Deciduous and mixed forests, woodlands, parks, and suburban areas with mature trees. Prefers a dense canopy and open understory for hunting insects.

Feeds on insects caught midair or picked off foliage. Perches on branches or exposed limbs and launches short flights to catch prey. Distinctive *pee-a-wee* call serves as territorial display and communication.

Eastern Wood-Pewees are known for their sallying hunting technique, where they dart out from a perch to catch insects midair then return to the same perch.

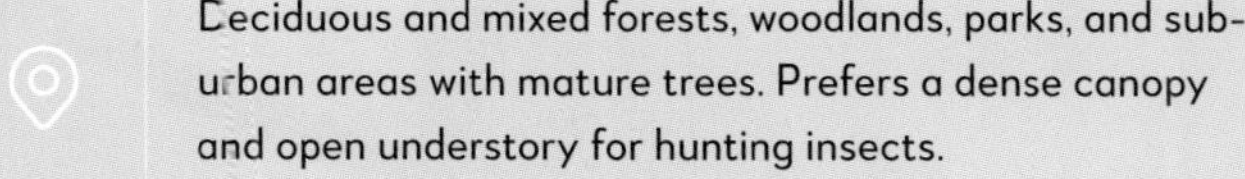

LENGTH: 5" / WINGSPAN: 9–10"

Adult.
Small. Peaked head with eyering. Olive-gray body with wingbars.

LEAST FLYCATCHER

Empidonax minimus

Small flycatcher with olive-gray upperparts, pale yellowish belly, and distinct eyering. Relatively short bill and compact size compared to other flycatcher species.

Forests, woodlands, and mixed deciduous-coniferous forests. Prefers dense understory vegetation, often near streams or wetlands.

Feeds on small flying insects such as flies, mosquitoes, and beetles. Perches on branches or in understory and makes quick sallies to capture prey.

The Least Flycatcher is known for its call, a sharp *che-bec* repeated several times. Males use this vocalization to attract mates and establish territories during breeding season.

LENGTH: 4–5" / WINGSPAN: 7"

Adult.

Dark head and face. Light throat and belly. Slightly peaked head with flattened bill.

EASTERN PHOEBE

Sayornis phoebe

Plump flycatcher with big head, dark bill, drab brownish-gray plumage, and light-colored underbelly.

Woodlands, forests, parks, and suburban areas with suitable perches and nesting sites. Often near water sources like streams, ponds, and marshes.

Feeds on insects caught midair or picked off foliage. Perches up high and waits before sallying out to catch a flying insect. *Phoebe* call is a territorial display and communication tool.

Eastern Phoebes frequently wag their tail while perched, a behavior observed in many flycatchers. This tail wagging may help them maintain balance and attract insects.

LENGTH: 5–6" / WINGSPAN: 10–11"

Male.
Electric-blue head and back. Rusty orange throat, breast, and sides.

Female.
Faded grayish-blue head and back. Rusty orange throat, breast, and sides.

EASTERN BLUEBIRD

Sialia sialis

Male with bright blue back and wings, rusty orange chest and belly. Female with more muted colors. Thin, straight bill.

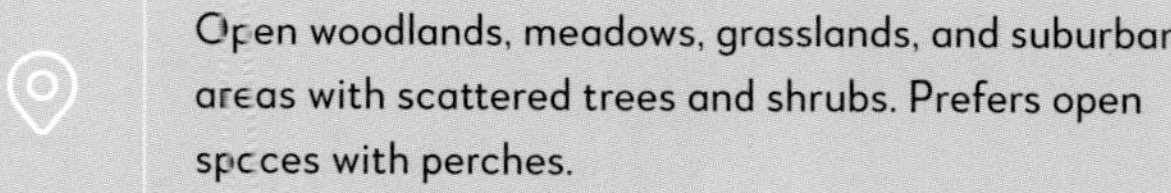

Open woodlands, meadows, grasslands, and suburban areas with scattered trees and shrubs. Prefers open spaces with perches.

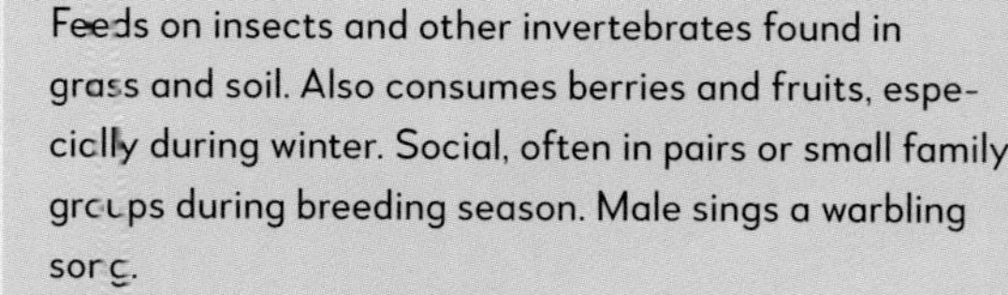

Feeds on insects and other invertebrates found in grass and soil. Also consumes berries and fruits, especially during winter. Social, often in pairs or small family groups during breeding season. Male sings a warbling song.

Eastern Bluebirds are cavity nesters but will use human-made nest boxes. Conservation efforts such as nest box installation has helped increase bluebird populations across the region.

LENGTH: 6–8" / WINGSPAN: 9–12"

Adult.
Plain, olive-brown upperparts. Creamy white underbelly with faint spots.

GRAY-CHEEKED THRUSH

Catharus minimus

Plain, olive-brown thrush with faintly spotted, creamy white underbelly. Distinct gray patch on cheek.

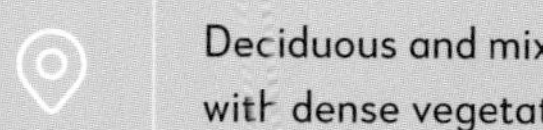

Deciduous and mixed woodlands, parks, and gardens with dense vegetation.

Feeds on insects, spiders, and other invertebrates. Hops on ground or in low vegetation, probing leaf litter and foliage for prey. Uses ethereal song to establish territory and attract mate. Forms monogamous pairs. Long-distance migrant between North American breeding grounds and South American wintering grounds.

The Gray-cheeked Thrush is often heard but not easily spotted, adding an element of mystery to birdwatching adventures during migration.

LENGTH: 6" / WINGSPAN: 12–13"

Adult.
Warm brown upperparts. Large eye. Brown spots on underbelly.

SWAINSON'S THRUSH

Catharus ustulatus

Thrush with warm brownish-olive upper plumage, creamy white underbelly with brown spots, and distinctive buffy eyering.

Coniferous, deciduous, and mixed woodlands, parks, and gardens with dense vegetation.

Feeds on insects, spiders, and other invertebrates picked off ground and from leaf litter. Also eats fruit during nonbreeding season. Sings unforgettable melodious and flute-like song during breeding to establish territory and attract mate.

Swainson's Thrushes migrate between the Americas from their breeding grounds to their wintering grounds, traveling at night and using celestial cues for navigation.

LENGTH: 6–7" / WINGSPAN: 11–12"

Adult.
Thin bill. Rusty red tail. Spotting on chest.

HERMIT THRUSH

Catharus guttatus

Rich brown thrush with eye-catching reddish tail. White underparts with smudged dark spots. Pink legs and thin white eyering.

Dense deciduous and mixed woodlands with ample understory vegetation. Also wooded suburban areas, parks, and gardens with suitable cover and food sources.

Feeds on insects, spiders, and small invertebrates picked off ground and from leaf litter. Also consumes berries, fruits, and seeds, particularly in fall and winter. Solitary, often foraging quietly on forest floor. Sings unmistakable melodious flute-like song.

Hermit Thrushes rustle the grass with their feet to uncover hidden insects and also quiver their feet after spotting a flying predator.

LENGTH: 5–7" / WINGSPAN: 9–11"

Male.
Black head. White around eye. Yellow bill. Rusty red breast.

AMERICAN ROBIN

Turdus migratorius

Earthy brown back and orange belly. Blackish head with white eyering and yellow bill. Male's orange belly and black head usually more vibrant than female's.

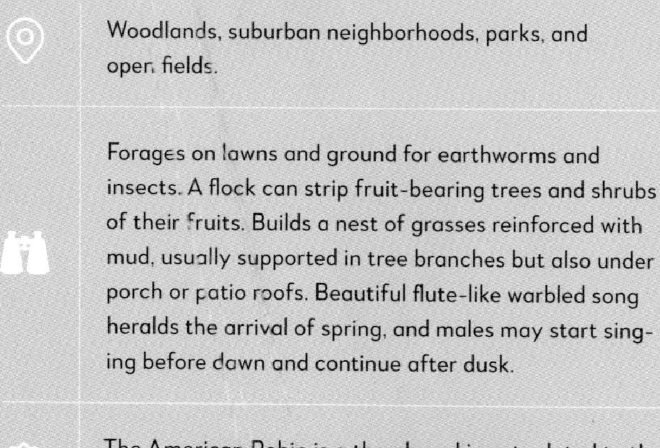

Woodlands, suburban neighborhoods, parks, and open fields.

Forages on lawns and ground for earthworms and insects. A flock can strip fruit-bearing trees and shrubs of their fruits. Builds a nest of grasses reinforced with mud, usually supported in tree branches but also under porch or patio roofs. Beautiful flute-like warbled song heralds the arrival of spring, and males may start singing before dawn and continue after dusk.

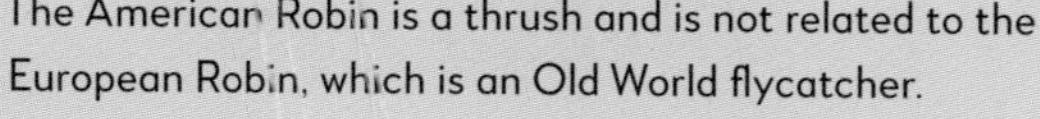

The American Robin is a thrush and is not related to the European Robin, which is an Old World flycatcher.

LENGTH: 7–11" / WINGSPAN: 12–15"

Adults.
Silky brown body, black mask, crested head. Secondary feathers and tail feathers tipped with bright colors.

CEDAR WAXWING

Bombycilla cedrorum

Silky-looking light brown bird with gray and brown wings, black mask, and crested head. Tail feathers tipped in bright yellow, and secondary flight feathers tipped in bright red.

Riparian areas, woodlands, orchards, parks, suburban areas; anywhere abundant fruit-bearing trees and shrubs are found.

Primarily consumes fruit but is acrobatic aerialist that catches flying insects. Highly social, often in flocks, especially during migration.

Cedar Waxwings will pass berries or small fruits from one bird to another before finally swallowing them. This "gifting" behavior is believed to strengthen social bonds within the flock.

LENGTH: 5–6" / WINGSPAN: 8–11"

Adult.
Gray face with black crown.

Adult.
Smooth medium-gray plumage with red undertail feathers.

GRAY CATBIRD

Dumetella carolinensis

Medium-sized songbird with uniform gray plumage, darker gray cap, rust-colored undertail feathers, black eyes, and inconspicuous black bill.

Thickets, shrubby areas, and forest edges with dense vegetation and tangled vines.

Feeds on insects, spiders, and other small invertebrates. Also consumes berries, fruits, and seeds, especially during fall and winter.

Gray Catbirds can mimic a variety of sounds, including the calls of other bird species, insects, and even mechanical noises.

LENGTH: 8–9" / WINGSPAN: 8–11"

Adult.
Long tail. Reddish-brown upperparts, heavily streaked underparts.

BROWN THRASHER

Toxostoma rufum

Medium-sized songbird with rich reddish-brown upper plumage, pale, heavily streaked underbelly. Striking yellow eyes, long, slightly downcurved bill, two prominent white wing bars.

Dense shrublands, thickets, forest edges, and brushy fields during summer. Open woodlands and suburban areas with ample shrubbery during migration.

Secretive and elusive ground forager. Flips leaves and digs in soil for insects, berries, and seeds. Impressive mimic with over 1100 songs in its repertoire.

Brown Thrashers are fiercely territorial during the breeding season. They defend their territory from other birds, humans, and small mammals.

LENGTH: 9–12" / WINGSPAN: 11–12"

Adult.
Light gray all over; black-and-white wings. Long tail.

NORTHERN MOCKINGBIRD

Mimus polyglottos

Slender bird with long tail, grayish-brown upper plumage, whitish-grey underbelly, and prominent white wingbars and patches.

Urban and suburban areas, parks, gardens, shrubby habitats and open woodlands with scattered trees. Favors grassy areas.

Feeds on insects, spiders, fruits, berries, and seeds. Assertive and adaptable. Dynamic forager, hopping and scratching ground to uncover prey and catch flying insects. Often perches on fences, wires, or treetops while singing.

Northern Mockingbirds are renowned for their extraordinary talent in mimicking the songs of other bird species and even sounds from their surroundings. They can accurately imitate several species and add these tunes to their vast collection of songs and vocalizations.

LENGTH: 8–10" / WINGSPAN: 12–13"

Breeding.
Glossy black plumage. Yellow, sharply pointed bill.

Nonbreeding.
White arrows across black plumage. Dark, sharply pointed bill.

EUROPEAN STARLING

Sturnus vulgaris

Medium-sized songbird with glossy black plumage that shows purple and green iridescence during the breeding season. Its yellow bill is short and straight.

Urban and suburban areas, agricultural fields, grasslands, and open woodlands. Thrives around human habitation.

Feeds on insects, fruits, seeds, and scraps on the ground, in trees, and shrubs, using sharp bill to probe and pluck food. Often in large flocks with noisy vocalizations.

European Starlings are talented mimics capable of imitating the calls of up to 20 bird species. They're truly the impressionists of the bird world, with a repertoire that can include over 60 songs!

LENGTH: 7–9" / WINGSPAN: 12–15"

Male.

Red wash on face and chest. Streaked belly, brown back. Sturdy conical bill.

Female.

Overall brownish gray with blurry streaks. Sturdy conical bill.

HOUSE FINCH

Haemorhous mexicanus

Male with bright red head, throat, and chest. Female with subdued brown and gray faintly streaked feathers. Sturdy conical bill.

Urban areas, suburban neighborhoods, parks, gardens, and open woodlands. Attracted to bird feeders and fruit-bearing trees and shrubs.

Feeds on seeds, grains, fruits, and insects. Often sings while perched on a branch or power line.

The bright red color seen in male House Finches is from carotenoid pigments they absorb from their food during their molting period.

LENGTH: 5" / WINGSPAN: 7–9"

Female.
Streaked body, red cap. Male's chest washed in pink.

REDPOLL

Acanthis flammea

Small finch with streaked brown back and wings, reddish cap, black chin. Pink wash on male's chest. Flanks streaked. Stubby, conical bill.

Open habitat such as fields, scrublands, and forests with dense vegetation.

Winter flocks extract seeds from birch and alder catkins, but also from grasses and other trees and shrubs. Highly social, giving cheerful chirps and trills within flock.

Redpolls form loose colonies in the Arctic tundra during the breeding season. They build cup-shaped nests in low shrubs or on the ground, lining them with feathers, moss, and plant fibers for their eggs and young.

LENGTH: 4–5" / WINGSPAN: 7–8"

Breeding male.
Brilliant yellow with black-and-white wings and tail, black forehead. Pink bill.

Nonbreeding male.
Brownish yellow with black patterned wings.

AMERICAN GOLDFINCH

Spinus tristis

Breeding male iconic bright yellow with black cap, black-and-white wings, pink conical bill. Female greenish yellow with black-and-white wings, pinkish conical bill.

Fields, grasslands, meadows, and open woodlands. Flocks particularly attracted to areas with abundant seed sources.

Feeds on seeds extracted from seed heads in weedy fields, especially thistle and sunflower. Visits bird feeders for seed. Performs acrobatic flights, accompanied by musical trills and twittering notes.

American Goldfinches undergo a seasonal transformation, molting bright yellow plumage for a more subdued olive hue during winter. This adaptation helps them blend into their environment and survive harsh winter conditions while maintaining their charming demeanor.

LENGTH: 4–5" / WINGSPAN: 7–8"

Adult.
Rust-colored crown, white eyebrow, black eyeline. Gray breast.

CHIPPING SPARROW

Spizella passerina

Small songbird with rust-colored crown, gray face, white eyebrow, and dark eyeline. Brown back with dark stripes, and soft gray underparts.

Open woodlands, scrublands, parks, gardens, and suburban areas.

Feeds on a variety of seeds, grains, and grasses. Highly social, often in small to medium-sized flocks during nonbreeding season. Song is high-pitched, one-note, repeated trill.

During the chilly months, the Chipping Sparrow transforms into a seedeating machine, consuming 70 times its weight in seeds to stay nourished, warm, and energized.

LENGTH: 4–5" / WINGSPAN: 8"

Adult.
Bold eyebrow, dark eye-line, thin dark mustache borders white stripe.

Adult.
Unstreaked belly. Blends in with ground.

CLAY-COLORED SPARROW

Spizelia pallida

Small songbird with whitish stripe above eye, brown crown bordered by dark stripes, and gray nape. Streaked brown upper body, plain buffy-white underparts. May have indistinct wingbars.

Open grasslands, prairies, agricultural fields, and field edges.

Feeds on seeds, grains, and grasses, especially during nonbreeding season. Forages by hopping on the ground, perching on low branches, and searching for food. Buzzy song often described as rapid, insect-like buzzes or trills.

During breeding, Clay-colored Sparrows form seasonal monogamous bonds but often switch partners yearly. Males defend territories for potential reunions, while females seek new nesting sites, making each breeding season a dynamic cycle of partnerships.

LENGTH: 4–5" / WINGSPAN: 7–8"

Adult.
Brownish-gray all over. Reddish-brown crown. Pale pink or gray bill.

FIELD SPARROW

Spizella pusilla

Small songbird with warm brownish-gray back and soft buff-colored underparts. Reddish-brown crown, white eyering, pale pink or gray bill, and light wingbars.

Grasslands, meadows, pastures, and agricultural fields. Prefers tall grasses and dense vegetation.

Feeds on ground for grass seeds, weed seeds, and grains, but switches gears during breeding season and consumes insects and protein-rich food.

The Field Sparrow sings a speedy song resembling the sound of a bouncing ping-pong ball. This playful and melodious tune adds a charming touch to the season's symphony of sounds.

LENGTH: 4–5" / WINGSPAN: 7"

Adult.
Rich reddish-brown cheek, back, and streaking. Gray above eye and on nape. Yellowish bill.

FOX SPARROW

Passerella iliaca

Medium-sized sparrow with stout yellow bill, rich warm brown or reddish-brown plumage, and intricate streaking and spotting on back and sides. Breast pale, buff-colored with bold, dark streaks.

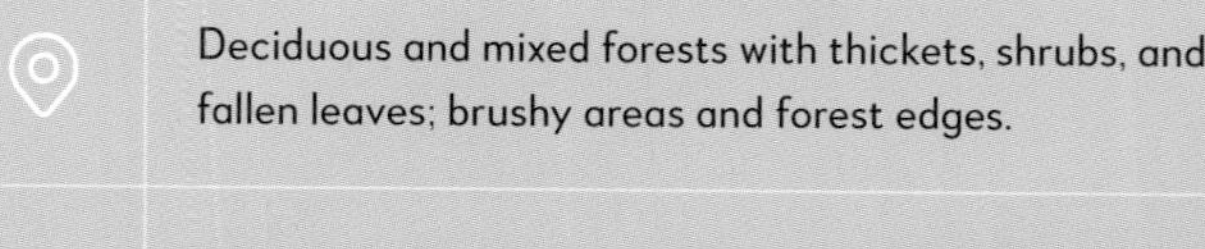

Deciduous and mixed forests with thickets, shrubs, and fallen leaves; brushy areas and forest edges.

Feeds on insects, seeds, and small invertebrates. Uses strong legs and sturdy bill to scratch and dig through leaf litter. Also consumes berries and fruits, especially during winter when insect prey is scarce. Musical and varied song consists of clear flute-like notes, trills, and warbles and serves to attract mates and communicate with other sparrows.

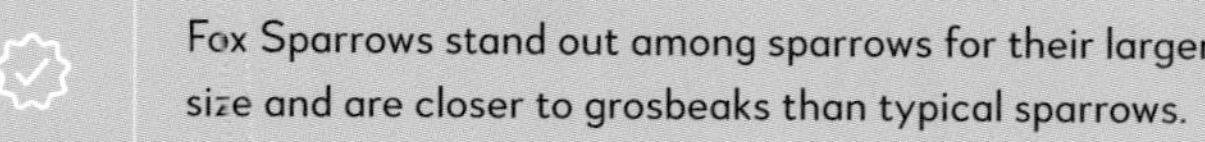

Fox Sparrows stand out among sparrows for their larger size and are closer to grosbeaks than typical sparrows.

LENGTH: 5–7" / WINGSPAN: 10–11"

Adult.
Russet cap and eyeline. Gray face. Bicolor bill.

AMERICAN TREE SPARROW

Spizelloides arborea

Small songbird with russet cap and eyeline, gray face, and bicolor bill. Dark spot on pale breast.

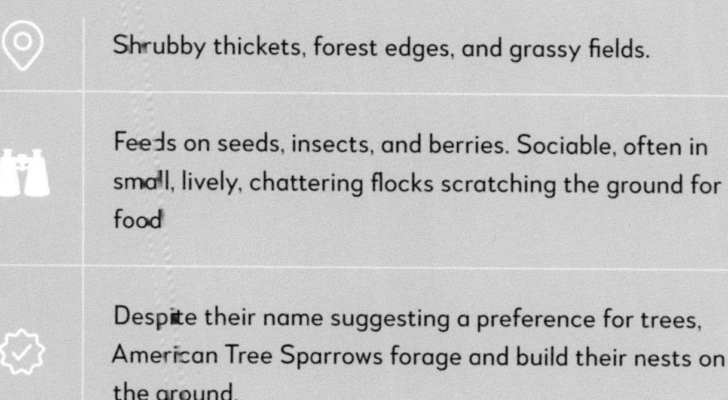

Shrubby thickets, forest edges, and grassy fields.

Feeds on seeds, insects, and berries. Sociable, often in small, lively, chattering flocks scratching the ground for food

Despite their name suggesting a preference for trees, American Tree Sparrows forage and build their nests on the ground.

LENGTH: 5" / WINGSPAN: 9"

Adult.
Gray hood, light bill. White belly.
White outer tail feathers.

DARK-EYED JUNCO

Junco hyemalis

Small songbird with slate-gray or brownish-gray back and wings, white belly and underparts. Black or dark gray hood, and pale pinkish bill.

Forests, woodlands, suburban areas, and open fields. Prefers dense shrubbery, brush piles, and leaf litter.

Forages on ground for seeds, grains, and grasses. Highly social, often in flocks. Communicates with tinkling trill calls.

Dark-eyed Juncos appear as winter arrives in the Great Lakes region. As spring sets in, they retreat northward, a seasonal migration pattern that aligns with changing weather and habitats.

LENGTH: 5–6" / WINGSPAN: 7–9"

Adult.
Striking black-and-white crown.

First winter.
Crown is brown and tan instead of black and white as in adult.

WHITE-CROWNED SPARROW

Zonotrichia leucophrys

Adult with boldly black-and-white striped head, gray face, brown wings, and gray breast. Juvenile with less distinct head pattern and overall browner.

Shrubby areas, brushy fields, dense shrubs, and edges of forests.

Feeds on seeds but also consumes insects, especially during breeding. Forages on ground, scratching and pecking through leaf litter. During breeding season, sings sweet song composed of clear whistled notes.

The scientific name *Zonotrichia leucophrys* translates from ancient Greek as "white eyebrow," referring to the white stripe above its eye.

LENGTH: 5–6" / WINGSPAN: 8–9"

Nonbreeding.
Warm buff face. Pink bill. Blackish crown. White belly.

HARRIS'S SPARROW

Zonotrichia querula

Large sparrow with blackish crown and bib, warm buff face, pink bill, white belly. Long tail.

Hedgerows, agricultural fields, shrubby pastures, backyards, and shrubby areas near streams. Present in winter.

Feeds on seeds, grains, and grasses. Forages by hopping on ground, perching on low branches, and searching for seeds and grains.

Harris's Sparrow is the only songbird in North America known to breed solely in Canada, making it an exceptional inhabitant of northern territories.

LENGTH: 6–7" / WINGSPAN: 10"

Adult.
Yellow between bill and eye. Crisp white throat and crown. Gray body, brown-and-black wings.

WHITE-THROATED SPARROW

Zonotrichia albicollis

Boldly striped black-and-white crown, bright white throat outlined in black, and yellow spot in front of each eye. Upperparts reddish brown with dark streaks, and underparts pale gray with faint dark streaks.

Coniferous and deciduous forests and mixed woodlands with thick undergrowth; shrubs near forest borders.

Feeds on seeds, insects, and small invertebrates found in leaf litter and on ground. Also consumes fruits and berries, especially during nonbreeding season.

Their *oh-sweet-Canada-Canada-Canada* song resonates through the woodlands, making it a characteristic sound of the region.

LENGTH: 6–7" / WINGSPAN: 7–9"

Adult.
Light brown back and wings, dark streaking on chest and sides. Yellow eyebrow.

SAVANNAH SPARROW

Passerculus sandwichensis

Light-colored sparrow with streaked brown upper body, delicately streaked pale breast, white belly. Short tail, small, conical bill. Yellow eyebrow.

Open grasslands and meadows, especially near wetlands or bodies of water. Also agricultural fields and abandoned farmland.

Feeds on insects, spiders, and seeds. Buzzy song sung from prominent perches to establish territory and attract potential mates during breeding season.

The Savannah Sparrow was named for Savannah, Georgia, where one of the first specimens was gathered for research purposes.

LENGTH: 4–5" / WINGSPAN: 7–8"

Adult.
Brown crown, gray eyebrow, brown eyeline, gray cheek.

SONG SPARROW

Melospiza melodia

Brown sparrow with gray and reddish-brown striped head, white throat, and streaked breast and belly. Dark spot in center of chest.

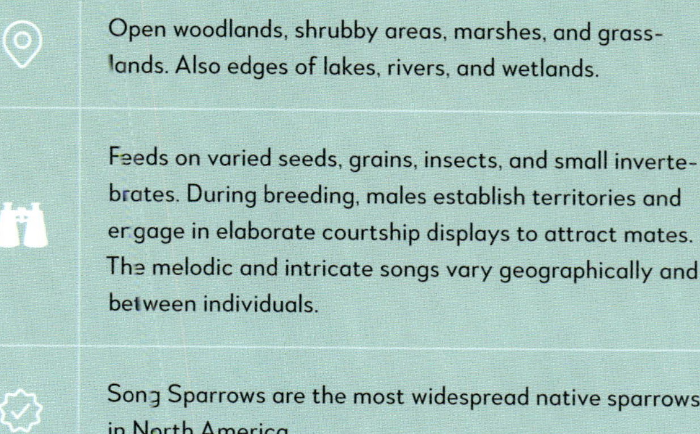

Open woodlands, shrubby areas, marshes, and grasslands. Also edges of lakes, rivers, and wetlands.

Feeds on varied seeds, grains, insects, and small invertebrates. During breeding, males establish territories and engage in elaborate courtship displays to attract mates. The melodic and intricate songs vary geographically and between individuals.

Song Sparrows are the most widespread native sparrows in North America.

LENGTH: 4–6" / WINGSPAN: 7–9"

Adult.
Fine streaks on buffy chest and sides. Pale eyering.

LINCOLN'S SPARROW

Melospiza lincolnii

Small sparrow with streaked brown back, buffy breast with fine streaks, white belly, distinct pale eyering, and small bill.

Dense, moist shrubby areas, and brushy thickets with ample ground cover near marshes, stream edges, and wet meadows.

Feeds on insects, spiders, and arthropods in leaf litter and dense vegetation. Secretive and tends to stay in cover. Forages by hopping on ground and scratching through debris.

Lincoln's Sparrows are known for their subtle and soft tinkling song, distinct from the more robust songs of other sparrows.

LENGTH: 5" / WINGSPAN: 7–8"

Adult.
Rust-colored cap tops a gray face.

SWAMP SPARROW

Melospiza georgiana

Sparrow with reddish-brown back and wings, gray face with white throat, and rust-colored cap extending down nape. Relatively short, conical bill and rounded tail.

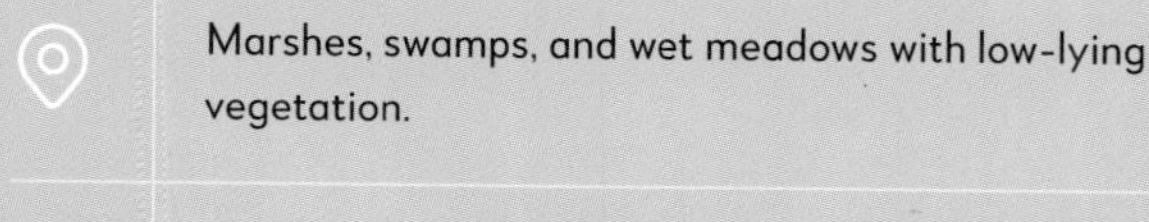

Marshes, swamps, and wet meadows with low-lying vegetation.

Feeds on seeds, fruits, and invertebrates. Forages on or near ground. Probes with bill, pecks at vegetation, and kicks up leaf litter to uncover prey. During breeding, males establish territories and perform elaborate singing displays to attract mates, while females construct cup-shaped nests on or near ground.

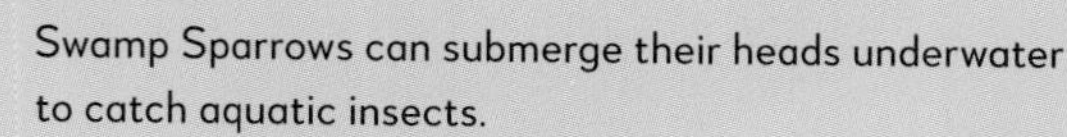

Swamp Sparrows can submerge their heads underwater to catch aquatic insects.

LENGTH: 4–5" / WINGSPAN: 7"

Male.
Jet-black head and chest, red eye. Rufous sides, white belly.

EASTERN TOWHEE

Pipilo erythrophthalmus

Large sparrow with predominantly black body, long tail with white spots, rich rufous sides, and white belly. Black wings, bright red eyes.

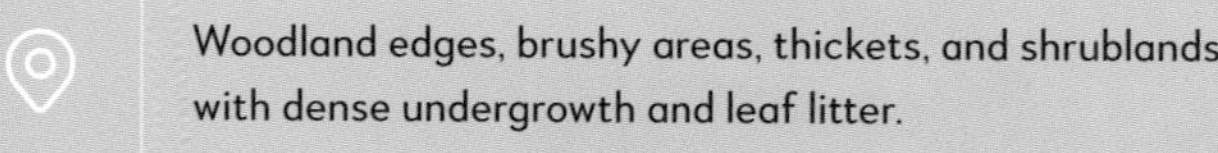

Woodland edges, brushy areas, thickets, and shrublands with dense undergrowth and leaf litter.

Feeds on insects, seeds, and small invertebrates. Forages on ground, using strong legs and bill to scratch and dig through leaf litter. Also consumes berries and fruits, especially during winter. *Drink-your-tea* song used for territorial display and communication.

Eastern Towhees are known for their powerful double scratch, where they quickly scratch the ground forward and then back to uncover food.

LENGTH: 6–8" / WINGSPAN: 7–11"

Male.
Gray crown, black throat and bib. Gray body, brown wings.

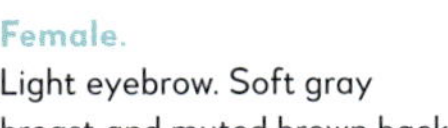

Female.
Light eyebrow. Soft gray breast and muted brown back.

HOUSE SPARROW

Passer domesticus

Sparrow with brown, gray, and black plumage. Males have black bib, and females are more subdued. Larger bill than typical native sparrows.

Urban and suburban areas, agricultural lands, and open woodlands with human habitation nearby.

Feeds on seeds, grains, and small insects. Also consumes fruits and berries. Adaptable and social. Often perches on buildings, fences, or power lines and forages on the ground and in vegetation. Often cheeps loudly.

House Sparrows closely associate with humans, often nesting in and around buildings, barns, and other structures. They have adapted to urban and suburban environments, making them familiar sights at backyard feeders and throughout cityscapes.

LENGTH: 5–6" / WINGSPAN: 7–9"

Nonbreeding.
White plumage with reddish accents; black in wings. Yellowish bill.

Nonbreeding.
Reddish band on chest.

SNOW BUNTING

Plectrophenax nivalis

Sparrow-like bird, mostly white, with rust-colored patches on head and chest. Small conical bill turns yellow in nonbreeding season, adding color to winter plumage.

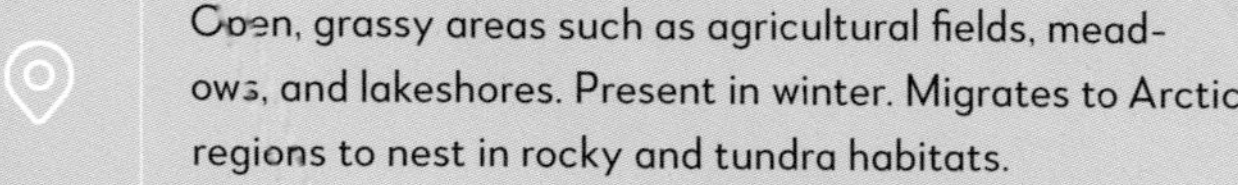

Open, grassy areas such as agricultural fields, meadows, and lakeshores. Present in winter. Migrates to Arctic regions to nest in rocky and tundra habitats.

Feeds on seeds, but also consumes insects and small invertebrates. Highly social, particularly during migration and winter, forming large flocks and performing acrobatic flight displays and synchronized movements.

Snow Buntings undergo seasonal changes in plumage, transitioning from mostly rust colored and white with rusty orange accents during the nonbreeding season to white with black wings in the breeding season. Their white plumage helps camouflage them in their preferred arctic breeding habitat.

LENGTH: 5" / WINGSPAN: 11"

Female.
Yellow wash over face and chest. Brown back and tail.

Male.
Bright yellow head. Black body with white wing patches.

YELLOW-HEADED BLACKBIRD

Xanthocephalus xanthocephalus

Bright yellow head and throat of male contrasts with black body and wings. Large white patch on wing. Female and juvenile have splotchy yellow face on dark brown body.

Marshes, reed beds, and grasslands adjacent to water bodies.

Feeds on insects, spiders, and other small invertebrates, but also consumes seeds and grains, particularly during nonbreeding season. Breeds in loose colonies and constructs cup-shaped nests in dense vegetation near water. Parents share incubation and care for the young. Male's song is unusual blend of groans, whines, and creaks.

The male's bright yellow head and throat serve as a visual display during courtship and territorial defense.

LENGTH: 8–10" / WINGSPAN: 16–17"

Male.
Buff back of head; black face. Light rump.

BOBOLINK

Dolichonyx oryzivorus

Male a striking combination of black back, white rump, and buffy yellow nape. Female sparrow-like but with larger bill, warm buffy plumage, and dark line behind eye.

Grasslands, meadows, and hayfields with tall grasses and scattered shrubs. Also agricultural landscapes with a mix of open fields and shrubby cover.

Feeds on insects and seeds. Male's mechanical, bubbly songs delivered in rapid succession. Forms monogamous pairs and builds cup-shaped nest in tall grasses or dense vegetation near ground.

One of its nicknames is "skunk blackbird," inspired by the male's striking black-and-white breeding plumage. Its colors are skunk-like, but it has a lively touch of yellow on its head.

LENGTH: 5–8" / WINGSPAN: 10"

Male.
Black head and back, orange belly. Sharply pointed bill.

Female.
Yellow all over, with sharply pointed bill.

BALTIMORE ORIOLE

Icterus galbula

Male striking inky black with orange belly and pointed bill. Female softer yellow and orange tones, also with pointed bill.

Habitats with mature trees, such as deciduous forests, orchards, and suburban areas.

Feeds on ripe berries, caterpillars, beetles, and grasshoppers. Acrobatic. Hangs upside down from branches while feeding. Weaves intricate hanging nest from grasses, plant fibers, and other materials.

During breeding season, males serenade the world with a rich, melodic, flute-like song, marking the start of spring.

LENGTH: 6–7" / WINGSPAN: 9–11"

Male.
All black with red shoulder patch bordered by yellow. Pointed bill.

Female.
Heavily streaked all over. Light eyebrow. Pointed bill.

RED-WINGED BLACKBIRD

Agelaius phoeniceus

Male glossy black with bright red-and-yellow epaulets, pointed bill. Female smaller, resembles large sparrow, with overall heavy streaking, paler throat, and pointed bill.

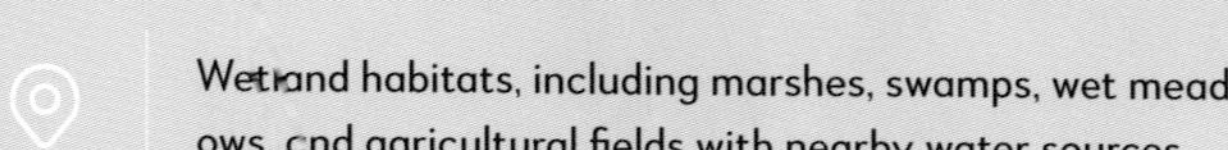

Wetland habitats, including marshes, swamps, wet meadows, and agricultural fields with nearby water sources.

Feeds on seeds, grains, insects, and small invertebrates. Males establish and defend breeding territories, displaying colorful red epaulets and singing frequently from cattails. Both sexes vocalize frequently. Highly social and forms large flocks outside breeding season, particularly during migration and winter.

Male Red-winged Blackbirds are polygynous and may mate with up to 15 different females during a breeding season.

LENGTH: 6–9" / WINGSPAN: 12–15"

Male.
Brown head, glossy black body. Conical bill.

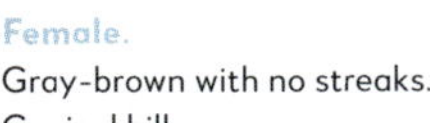

Female.
Gray-brown with no streaks. Conical bill.

BROWN-HEADED COWBIRD

Molothrus ater

Male medium-sized with glossy black plumage and matte brown head and neck. Female is slightly smaller and duller in color.

Grasslands, agricultural fields, and forest edges. Also areas with mix of open spaces and shrubby or wooded areas.

Feeds on seeds, grasshoppers, beetles, and other insects. Highly social, especially during nonbreeding season. Known for loud and varied vocalizations, including whistles, chirps, and gurgling calls, adding to the lively atmosphere of their habitat.

Brown-headed Cowbirds are brood parasites and lay their eggs in the nests of other bird species, relying on those hosts to raise their young. This behavior has made them controversial among conservationists due to the impact on host species populations.

LENGTH: 7–8" / WINGSPAN: 14"

Male.
Yellow eye. Sharply pointed bill. Iridescent blue, black, bronze plumage.

COMMON GRACKLE

Quiscalus quiscula

Large blackbird with iridescent black plumage that reflects metallic blue or purple hues. Female with less iridescent feathers. Both have large yellow eyes and long tail.

Urban and suburban areas, agricultural fields, woodlands, and marshes. Particularly abundant in open spaces near water sources.

Feeds on seeds, insects, small frogs and fish. Uses strong bill to crack shells, probe soil, or catch live prey. Highly social and gathers in large flocks or mixed species foraging groups, especially during nonbreeding season.

The male's iridescent head changes color depending on the angle of the light and can appear blue, purple, or green.

LENGTH: 11–13" / WINGSPAN: 14–18"

Adult.
Bold eyebrow. Brownish-olive upperparts, creamy underparts with dark streaking.

NORTHERN WATERTHRUSH

Parkesia noveboracensis

Small migratory songbird with brownish-olive upper plumage, heavily streaked creamy white underbelly, white eyebrow, and long legs.

Marshes, bogs, streams, and wooded swamps. Prefers dense vegetation and shallow water.

Feeds on insects and small invertebrates found in and around water. Active. Bobs tail while foraging among vegetation and leaf litter.

The Northern Waterthrush is known for "teetering," constantly bobbing its rear up and down as it walks. The motion helps the waterthrush locate and catch insects hidden in the vegetation and leaf litter.

LENGTH: 4–5" / WINGSPAN: 8–10"

Male.
Black throat, gold crown, white eyebrow and mustache, black eyeline. Gray body with yellow wing patches.

GOLDEN-WINGED WARBLER

Vermivora chrysoptera

Striking warbler with golden-yellow crown, black throat and mask. Gray body, yellow patches on wings.

Young forests, shrubby areas, and open woodlands such as regenerating clear-cuts, abandoned fields with scattered trees, and edges of wetlands.

Feeds on insects, caterpillars, and spiders in foliage. Also consumes fruits, berries, and seeds, especially during nonbreeding season. Flight is swift and direct.

The Golden-winged Warbler's insect-like *bee-buzz-buzz-buzz* song is a delightful sound often heard in the early mornings of their breeding season.

LENGTH: 5" / WINGSPAN: 7–8"

Male.
All over black-and-white striped. Black ear patch, white eyebrow and mustache.

BLACK-AND-WHITE WARBLER

Mniotilta varia

Warbler with striped black-and-white plumage and no yellow. Male has black ear patch, and both sexes have small, slightly curved bill.

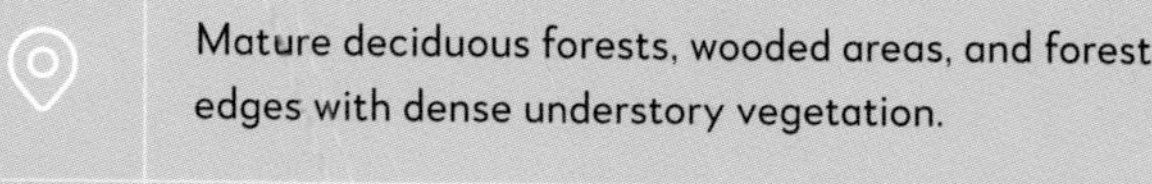

Mature deciduous forests, wooded areas, and forest edges with dense understory vegetation.

Feeds on insects, spiders, and other invertebrates found in bark crevices. Forages by creeping along tree trunks and branches much like nuthatch or Brown Creeper, using long, curved claws to grip and slender bill to probe for prey.

Black-and-white Warblers are among the earliest nesting warblers, nesting before many of their migratory companions arrive. This timing makes them one of the first warblers to appear in the Great Lakes region during spring migration.

LENGTH: 4–5" / WINGSPAN: 7–8"

Male.
Brilliant golden-yellow head and underparts. Thick black bill. Olive-green back, grayish wings.

PROTHONOTARY WARBLER

Protonotaria citrea

Very striking warbler with bright golden-yellow head and underparts contrasting with olive-green back and grayish wings. Thick black bill, blue-gray tail.

Bottomland forests, swampy areas, and wetlands with slow-moving or stagnant water, such as floodplain forests, marshes, and riverine habitats.

Feeds on insects, insect larvae, spiders, and other invertebrates. Nests in natural tree cavities or will adapt to artificial nest boxes. Known for loud, ringing song, which establishes territory, attracts a mate, and communicates with others within their social group.

Prothonotary Warblers are sometimes called "swamp canaries" because of their bright yellow plumage and preference for wetland habitats.

LENGTH: 5" / WINGSPAN: 8"

Male.
Gray head. Thin white eyering, pale eyebrow. Olive-green back and wings. Pale yellowish belly.

TENNESSEE WARBLER

Leiothlypis peregrina

Warbler with gray head, plain olive-green back and wings, and pale yellowish belly. Faint, thin white eyering, pale eyebrow, and short pointy bill. Females more yellow, which adds a touch of brightness to their appearance.

Deciduous and coniferous forests, woodlands, and open areas during migration. Also dense shrubs and undergrowth.

Feeds on caterpillars and other small invertebrates. Active and agile foragers, constantly flitting and hopping among branches and foliage, even hanging upside down or briefly hovering to capture prey.

Tennessee Warblers have a particular affinity for eating spruce budworms, causing their population to fluctuate with budworm outbreaks.

LENGTH: 3–5" / WINGSPAN: 7"

Male.
Gray hood, bold white eyering, yellow body. Rust-colored patch on crown.

NASHVILLE WARBLER

Leiothlypis ruficapilla

Bright yellow throat and underparts, olive-green back and wings. Gray head with white eyering, rust-colored patch on crown.

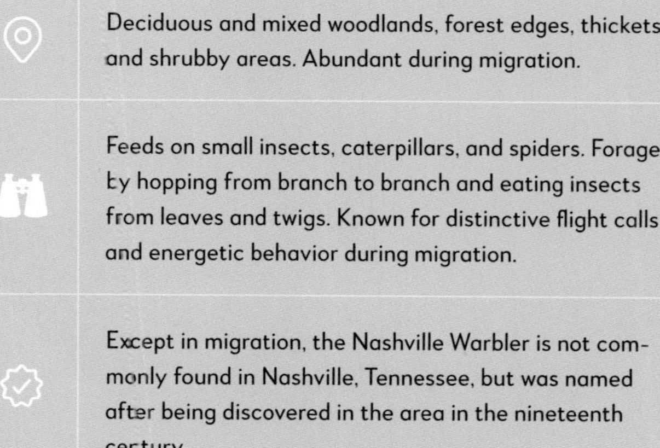

Deciduous and mixed woodlands, forest edges, thickets, and shrubby areas. Abundant during migration.

Feeds on small insects, caterpillars, and spiders. Forages by hopping from branch to branch and eating insects from leaves and twigs. Known for distinctive flight calls and energetic behavior during migration.

Except in migration, the Nashville Warbler is not commonly found in Nashville, Tennessee, but was named after being discovered in the area in the nineteenth century.

LENGTH: 4–5" / WINGSPAN: 6–7"

Male.
Black mask with white stripe above. Bright yellow throat and chest.

COMMON YELLOWTHROAT

Geothlypis trichas

Small songbird with bright yellow throat and breast, olive-green back and wings. Male has black mask across face.

Wetlands, marshes, shrubby areas, and pine forest edges. Prefers dense vegetation near open water.

Feeds on insects and invertebrates in dense vegetation. Active forager, hopping and flitting between branches while probing for prey.

Common Yellowthroats are highly vocal, known for their *wichety-wichety-wichety* song, which serves as a territorial signal and communication between mates and family members.

LENGTH: 4–5" / WINGSPAN: 5–7"

Male.

Black crown connected to black throat. Yellow face. Olive-green back and wings. Yellow belly.

HOODED WARBLER

Setophaga citrina

Male's striking black hood and throat contrast with bright yellow face and olive-green wings and back. This captivating appearance makes it a favorite among birdwatchers.

Mature deciduous forests, wooded areas, and forest edges with dense understory vegetation, particularly near streams and moist places.

Feeds on insects, spiders, and invertebrates. Forages by hopping and gleaning insects from leaves and twigs. Gives loud, ringing song during breeding season.

Male Hooded Warblers display territorial loyalty, with males returning each year to the same nesting area. In contrast, females explore and use new territories every breeding season.

LENGTH: 5" / WINGSPAN: 6"

Male.
Black head, chest, and back. Orange accents on upper chest, in wings and tail.

AMERICAN REDSTART

Setophaga ruticilla

Male is all black with bright orange patches on wings, tail, and sides. Female displays soft yellow patches instead.

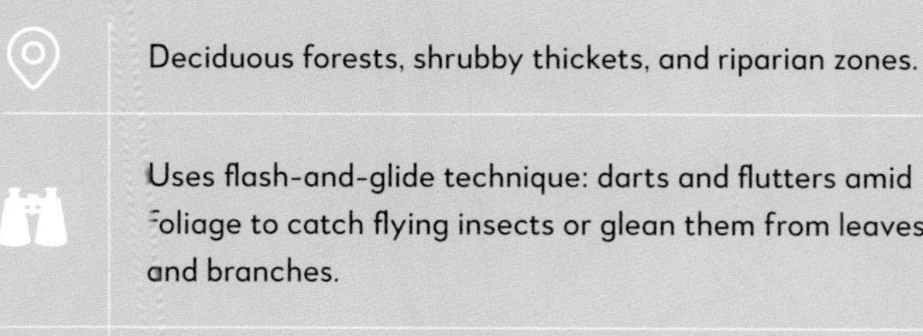

Deciduous forests, shrubby thickets, and riparian zones.

Uses flash-and-glide technique: darts and flutters amid foliage to catch flying insects or glean them from leaves and branches.

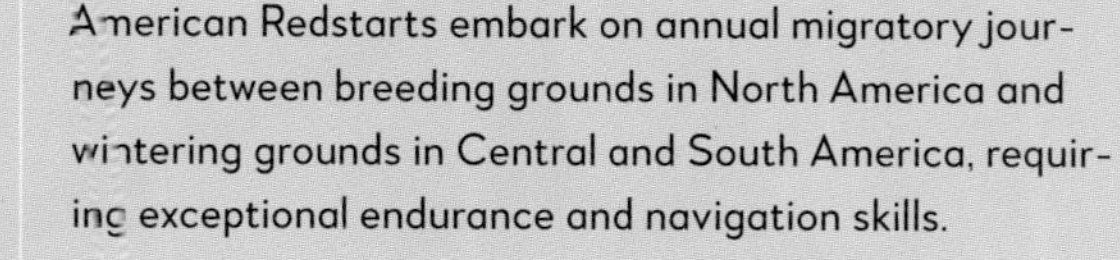

American Redstarts embark on annual migratory journeys between breeding grounds in North America and wintering grounds in Central and South America, requiring exceptional endurance and navigation skills.

LENGTH: 4–5" / WINGSPAN: 6–7"

Male.
Yellow throat and chest with chestnut-colored band, white eye crescents, bluish-gray back and face.

NORTHERN PARULA

Setophaga americana

Small, colorful warbler with bright yellow throat and underparts, bluish-gray back and wings. White eye crescents, yellow lower bill, dark chestnut-colored band across chest, and small orange patch on back.

Deciduous and mixed woodlands, swampy areas, and riparian zones. Prefers dense vegetation.

Feeds on small insects, caterpillars, and spiders. Active forager, often seen hopping from branch to branch and collecting insects from leaves and twigs.

The original name of the Northern Parula was Finch Creeper.

LENGTH: 4" / WINGSPAN: 6–7"

Male.
Yellow throat and belly with black streaking. Black mask, white eyebrow, gray cap and nape. White patch in wings.

MAGNOLIA WARBLER

Setophaga magnolia

Small colorful warbler with striking black mask across eyes, white eyebrow, and bright yellow throat and underparts. Black streaks on chest and flanks. Long, black-tipped tail.

Deciduous and mixed woodlands, forest edges, and shrubby areas. Abundant during migration.

Feeds on small insects, caterpillars, and spiders. Forages by hopping from branch to branch and gleaning insects from leaves and twigs. Known for distinctive flight calls and energetic behavior during migration.

During migration, Magnolia Warblers often team up with Black-capped Chickadees in foraging flocks, so listen for the cheerful *chick-a-dee-dee-dee* of the chickadee and watch for the colorful warbler.

LENGTH: 4–5" / WINGSPAN: 6–7"

Male.
Bright orange face and throat with black accents. Black wings with white patch.

BLACKBURNIAN WARBLER

Setophaga fusca

Male's fiery orange face and throat contrast against black-and-white back and white belly. One of the most sought-after and visually captivating species in the Great Lakes region.

Dense forests, particularly mature hemlock, spruce, fir, and white pine, for breeding. Deciduous forests during migration.

Feeds on insects gleaned from foliage or caught mid-air. Highly active and agile, flitting through canopy and probing foliage.

The Blackburnian Warbler is the only warbler in North America sporting an orange throat, making it easily recognizable and a delight for birdwatchers to spot.

LENGTH: 4" / WINGSPAN: 7–9"

Female and Male.
Yellow all over. Male with red streaks on chest.

YELLOW WARBLER

Setophaga petechia

Small, bright yellow songbird. Male has red or chestnut-colored streaks on breast and flanks. Females and juveniles pale yellow or greenish-yellow with faint streaks.

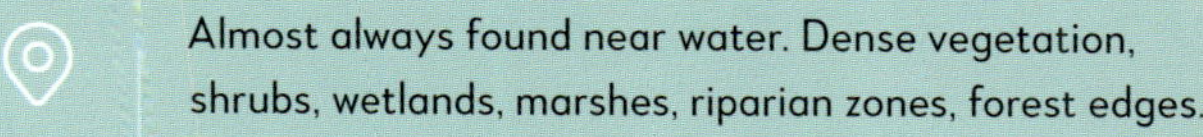

Almost always found near water. Dense vegetation, shrubs, wetlands, marshes, riparian zones, forest edges.

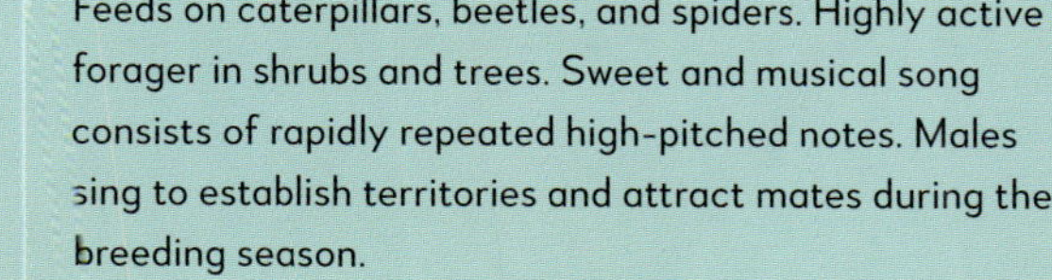

Feeds on caterpillars, beetles, and spiders. Highly active forager in shrubs and trees. Sweet and musical song consists of rapidly repeated high-pitched notes. Males sing to establish territories and attract mates during the breeding season.

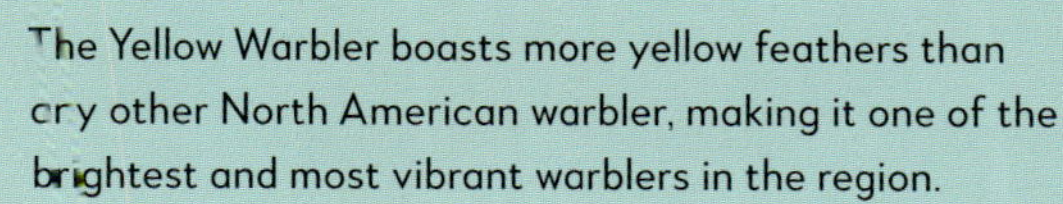

The Yellow Warbler boasts more yellow feathers than any other North American warbler, making it one of the brightest and most vibrant warblers in the region.

LENGTH: 4–5" / WINGSPAN: 6–8"

Male.
Chickadee-like face, with black crown and white cheek. Black streaks on back and flanks. Orange legs.

BLACKPOLL WARBLER

Setophaga striata

Male's black-and-white breeding plumage includes black cap, white cheeks, and streaked body, and no yellow. Female streaky with suggestion of black cap. Both genders have long orange legs, contributing to unique appearance.

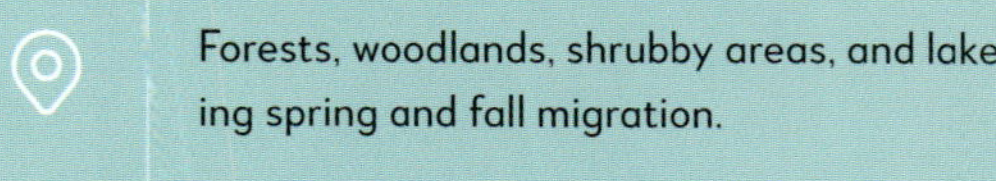

Forests, woodlands, shrubby areas, and lakeshores during spring and fall migration.

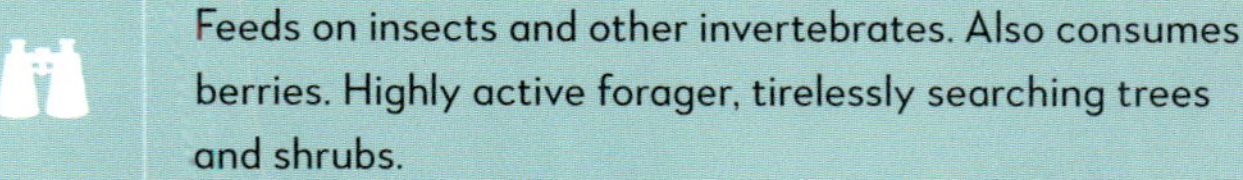

Feeds on insects and other invertebrates. Also consumes berries. Highly active forager, tirelessly searching trees and shrubs.

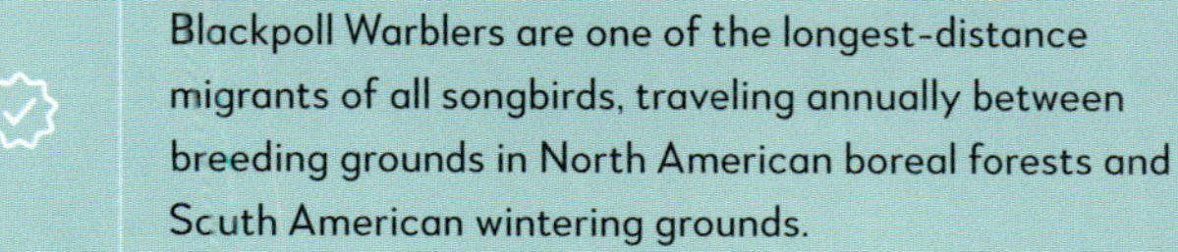

Blackpoll Warblers are one of the longest-distance migrants of all songbirds, traveling annually between breeding grounds in North American boreal forests and South American wintering grounds.

LENGTH: 5" / WINGSPAN: 8–9"

Male.
Rust-colored cap. Pale-yellow wash on throat. Rust-colored streaks on breast.

PALM WARBLER

Setophaga palmarum

Warbler with rust-colored cap, brownish-olive back with dark streaks, and yellow undertail coverts visible when tail is flicked. Pale-yellow wash on throat and rust-colored streaks on breast.

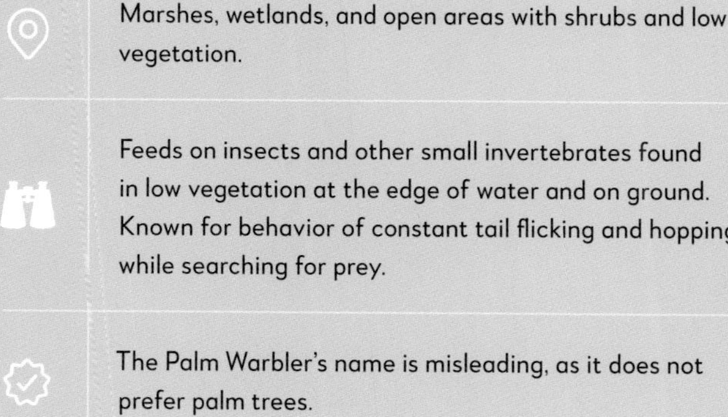

Marshes, wetlands, and open areas with shrubs and low vegetation.

Feeds on insects and other small invertebrates found in low vegetation at the edge of water and on ground. Known for behavior of constant tail flicking and hopping while searching for prey.

The Palm Warbler's name is misleading, as it does not prefer palm trees.

LENGTH: 4–5" / WINGSPAN: 7–8"

Male.
Yellow-olive upperparts and throat. Faint, broken eyering. Two white wingbars.

PINE WARBLER

Setophaga pinus

Warbler with yellow-olive upperparts and yellow throat. Two white wingbars; faint, broken yellow eyering; and thin, pointed bill.

Pine forests, mixed woodlands, and pine plantations. Prefers mature pine stands with an open understory.

Feeds on caterpillars, beetles, spiders, and other small insects found in pine trees and nearby vegetation. Melodious song consists of musical trills and warbles and is used to establish territory and attract a mate.

Pine Warblers prefer to nest in pine trees and use pine needles and other materials to build sturdy, well-concealed cup nests. These warblers fill pine forests with lively and cheerful songs, particularly in spring and summer.

LENGTH: 5" / WINGSPAN: 7–9"

The "butterbutt."

Male.
Black mask, white eye crescents, white throat and eyebrow. Yellow sides and rump. Gray body with black streaking.

YELLOW-RUMPED WARBLER

Setophaga coronata

Widespread warbler with gray back, white throat, and noticeable yellow on crown, sides, and rump. During breeding season, male displays bold black streaks on chest and flanks. "Myrtle" population of this species most common in region, but watch for "Audubon's," which has yellow throat, making it even more colorful.

Mixed forests, woodlands, and shrubby areas with diverse trees and dense vegetation. Coniferous forests during breeding; various habitats during migration and winter, including urban parks and gardens.

Feeds on insects, berries, fruits, and seeds. Highly adaptable forager. Active and agile, hops and flits among branches. Gives distinctive chip call and musical trill-like song, particularly during the breeding season.

They are affectionately nicknamed "butterbutt" because of the bright yellow patch above their tail.

LENGTH: 4–5" / WINGSPAN: 7–9"

Male.
Long bill. Black mask, white eyebrow. Yellow throat. Gray back, white in wings.

YELLOW-THROATED WARBLER

Setophaga dominica

Warbler with bold black face mask, black-and-white stripes on head and wings, and white eyebrow. Bright yellow throat and upper breast, and white underparts with black streaking.

Mix of trees and shrubs in mature deciduous forests, wooded areas, and pine forests near rivers, streams, and wetlands.

Gleans insects from tree bark, pine cones, and needles. Agile and adaptable forager. Clear and melodic song consists of rich, whistled notes used for territorial defense, mate attraction, and communication within social group.

Yellow-throated Warblers forage by methodically moving up and down tree limbs and trunks, a style reminiscent of Brown Creepers and Black-and-white Warblers.

LENGTH: 5" / WINGSPAN: 8"

Male.
Tiny. Olive-green upperparts, bright yellow underparts. Distinct, jet-black cap.

Female.
Colors slightly more muted than male. Fainter cap.

WILSON'S WARBLER

Cardellina pusilla

Small, bright yellow warbler with olive-green upperparts, bright yellow underparts, and jet-black cap on male. Female and juvenile more muted with fainter cap.

Shrubby areas, young forests, and riparian zones with thick vegetation. Deciduous and mixed woodlands, especially along the edges of streams, rivers, and wetlands.

Feeds on insects, spiders, and other small creatures. Active forager that moves quickly through foliage and low vegetation. Gives short song of slightly accelerating notes, which adds to their lively presence.

This bright yellow bird is one of the smallest warblers in North America.

LENGTH: 3–4" / WINGSPAN: 5–6"

Male.
Scarlet head and back.
Black wings and tail.

SCARLET TANAGER

Piranga olivacea

Male is brilliant scarlet with black wings and tail. Nonbreeding male and female more subdued olive-yellow with paler underparts.

Mature deciduous forests, wooded areas, and forest edges. Prefers mix of large trees and shrubs, particularly oak-hickory forests, and riparian woodlands.

Feeds on insects, spiders, and other invertebrates gleaned from leaves and twigs, but also enjoys various fruits. Occasionally takes short flights to catch flying insects. Melodic song resembles that of American Robin but with sore-throat-like quality.

Male Scarlet Tanagers are known for their vibrant red breeding plumage, but in fall they undergo a transformation and molt into an olive-yellow color.

LENGTH: 6" / WINGSPAN: 9–11"

Male.
Iconic bright red plumage and bill; black face and throat. Crest.

Female.
Light brown plumage, red bill. Hints of red in tail and crest (this bird's crest is lowered).

NORTHERN CARDINAL

Cardinalis cardinalis

Male is bright red with black face and throat. Female is brown with touches of red on wings and tail. Both sexes have crested head and orange-red conical bill.

Woodlands, suburban areas, parks, and gardens. Prefers dense shrubbery and ample vegetation.

Feeds on seeds, grains, fruits, and insects. Known for their melodic song, one of the most recognizable in the region.

Male Northern Cardinals feed their mates during courtship and nesting, demonstrating a unique and caring behavior among songbirds.

LENGTH: 8–9" / WINGSPAN: 9–12"

Male.
Black head and back;
white in wings. Pale bill.
Rose-red on white chest.

Female.
Bold light eyebrow.
Grayish-brown head and back;
white in wings; white belly with
streaks. Juvenile male similar.

ROSE-BREASTED GROSBEAK

Pheucticus ludovicianus

Male has black head, wings, and back, prominent rose-red patch on white breast. Female has bold white eyebrow and streaked brown-and-white plumage. Both sexes have heavy conical bill perfect for cracking seeds.

Deciduous and mixed wet forests, woodland edges, parks, and gardens with dense vegetation. Prefers habitats with shrubs and tall trees.

Feeds on seeds and nuts, but also fruit and insects. Males melodic, rich, warbling song sometimes incorporates sounds from the environment.

The Rose-breasted Grosbeak and the Northern Cardinal belong to the same bird family.

LENGTH: 7–8" / WINGSPAN: 11–13"

Male.
Brilliant blue plumage; black in wings.

Female.
Brown, unstreaked plumage. Hint of blue in tail.

Male in the process of molting.

INDIGO BUNTING

Passerina cyanea

Male's vibrant blue plumage contrasts with female's subdued brown coloration.

Habitats with dense shrubs and thickets. Prefers open woodlands, brushy areas, meadows, and forest edges.

Feeds on seeds, grains, insects, berries, and fruits. Melodic songs delivered in rapid whistled notes, especially during breeding season when males attract mates and establish territories.

The male Indigo Bunting's brilliant plumage is not due to pigment but rather the structure of their feathers, which refract and scatter light to produce a vibrant blue color.

LENGTH: 4–5" / WINGSPAN: 7–8"

Adult.
Yellow eyebrow, gray nape; black on throat. Rusty red shoulders.

DICKCISSEL

Spiza americana

Small, migratory songbird with yellowish breast and belly streaked with brown, grayish-brown back, and black V-shaped bib on throat.

Open grasslands, agricultural fields, meadows, and prairies with tall grasses and scattered shrubs. Often found in pastures, abandoned fields, and roadside areas.

Feeds on seeds, grains, and grasses by hopping on ground or perching on grass stems.

Although the Dickcissel's song is short, its buzzing melody of *dick-dick-ceessa-ceessa* makes it easily recognizable and memorable.

LENGTH: 5–6" / WINGSPAN: 9–10"

A Call to Bird Joy

Birdwatching in the Great Lakes region is more than a hobby; it's a gateway to a deeper connection with nature. The sights and sounds of birds in their natural habitats awaken our senses and remind us of the beauty and diversity of life around us. Whether you're a seasoned birder or just beginning your birding journey, each observation brings a sense of wonder and appreciation for the natural world.

From the cheerful chirps of Tree Swallows to the haunting calls of the Common Loon, each bird leaves an indelible mark on our hearts, reminding us of the beauty and wonder that surrounds us every day.

Throughout this book, I've shared my passion for birdwatching, not just as a hobby but as a way of life—a lifestyle choice that brings fulfillment and connection with nature.

Together, we've explored tools and techniques for successful birdwatching and delved into the lives of over 150 bird species with captivating images and insightful details about their appearances, habitats, and behaviors.

As you close these pages, I urge you to carry forward the spirit of Bird Joy. Whether peering through binoculars in your backyard, strolling through a local park, or embarking on a birding adventure, let every sighting fill you with wonder and gratitude. Please share your experiences, inspire others to join your journey, and continue to marvel at the feathered wonders that grace the skies and land around us.

The world of birds is vast and ever-changing, offering endless opportunities for learning and discovery. As you continue your birdwatching endeavors, embrace every moment as a chance to expand your knowledge and deepen your understanding of bird behavior, habitats, and conservation. Take note of seasonal changes, migration patterns, and unique characteristics of each species you encounter. Every observation, whether common or rare, contributes to your growth as a birder and as a steward of the environment.

Thank you for allowing me to be your birding tour guide. May your days be extra birdy, filled with enchanting songs, vibrant colors, and the boundless wonder of birds, not just in the Great Lakes region but wherever your birding adventures take you. Keep looking up, and keep paying attention, for the birds are always there, ready to bring you joy and remind you of the magic of nature.

FURTHER RESOURCES

To enrich your birdwatching experience in the Great Lakes region, consider exploring the following resources:

Field Guides

Invest in comprehensive field guides specific to the region, such as *The Sibley Field Guide to Birds of Eastern North America* or *Peterson Field Guide to Birds of Eastern and Central North America*. These guides provide detailed information on bird species, including plumage variations, habitats, and behaviors.

Birding Apps

Utilize birding apps like Merlin Bird ID and eBird to enhance your identification skills, track sightings, and connect with a community of fellow birders. These apps offer real-time data, species alerts, and interactive features that make birdwatching more engaging and educational.

Local Birding Groups

Joining a local birding group or club, such as the BIPOC Birding Club of Wisconsin or the Audubon Society chapters, provides opportunities to learn from experienced birders, participate in group outings, and contribute to conservation efforts in your area. These communities foster camaraderie, knowledge sharing, and a deeper appreciation for birds and their habitats.

Nature Centers, Parks, and Wildlife Refuges

Visit nearby nature centers, parks, wildlife refuges, and birding hotspots within the Great Lakes region. These protected areas offer prime birdwatching opportunities, guided tours, and educational programs that allow you to immerse yourself in the region's natural beauty while observing birds in their native habitats.

Birdwatching in the Great Lakes region is a rewarding experience that invites us to slow down, observe, and connect with the natural world. As you embark on your birding adventures, embrace curiosity, practice patience, and always respect wildlife.

REFERENCES

All About Birds, AllAboutBirds.org. "Online Bird Guide, Bird ID Help, Life History, Bird Sounds from Cornell Lab." Accessed 2024.

Dunn, Jon L., and Jonathan Alderfer. *National Geographic Field Guide to the Birds of North America, 7th Ed.* Washington, DC: National Geographic, 2017.

Kaufman, Kenn. *Kaufman Field Guide to Birds of North America*. Boston: Mariner Books, 2005.

Peterson, Roger Tory. *Peterson Field Guide to Birds of Eastern and Central North America, 7th Ed.* Boston: Mariner Books, 2020.

Sibley, David Allen. *The Sibley Field Guide to Birds of Eastern North America*. New York: Knopf, 2016.

PHOTO CREDITS

All photos by the author except those noted here.

Page 50 (top), jamesvancouver/istock.com
Page 52 (top), SteveByland/istock.com
Page 54, Steve Samples/istock.com
Page 108, Joesboy/istock.com
Page 94 (top), KenCanning/istock.com
Page 110, Geral Corsi/istock.com
Page 244 (bottom), Sandi Smolker/istock.com
Page 250 (top), Liz Leyden/istock.com
Page 256 (top), BrianLasenby/istock.com
Page 294 (bottom), pchoui/istock.com
Page 310, Gerald Corsi/istock.com
Page 338 (top), Leon Gin/istock.com

Family Silhouettes

Geese, Swans & Ducks by Kristtaps/iStock.com
Loons by Rahul Kolgaonkar/pngitem.com
Grebes by Birchside/gograph.com
Cormorants by piranjya/iStock
Pelican from pngitem.com
Bitterns, Herons & Egrets by rangepuppies/iStock
Rails & Coots by iDrawSilhouettes/creativefabrica.com
Cranes by ArtistMiki/shutterstock.com
Stilts & Avocets by Marzolino/shutterstock.com
Plovers by Birchside/gograph.com
Sandpipers & Snipes by Birchside/gograph.com
Gulls & Terns by Birchside/gograph.com

Turkey by Save nature and wildlife/shutterstock.com
New World Vultures by iDrawSilhouettes/creativefabrica.com
Osprey / Hawks & Eagles by vadimmmus/iStock.com
Falcons by PetrP/shutterstock.com
Pigeons & Doves by DeCe X/iStock
Owls by thesilhouettequeen/123rf.com
Kingfishers by Birchside/gograph.com
Woodpeckers from pinclipart.com
Jays & Crows by VectorSpMan/shutterstock.com
Hummingbirds by mr.Timmi/shutterstock.com
Swallows by Ivana Kontic/shutterstock.com
Chickadees & Titmice by SilhouetteGarden.com
Nuthatches / Creepers by wectors/123rf.com
Wrens by Loveleen/stock.adobe.com
Vireos based on a photo by VJAnderson / Wiki-media Commons (used under a CCA-SA 4.0 International license)
Kinglets based on an illustration by Viktoria Karpunina/shutterstock.com
Flycatchers by Bahau/shutterstock.com
Thrushes by Jackie/cleanpng.com
Waxwings by Bob Comix/creazilla.com (used under a CCA 4.0 license)
Catbirds, Thrashers & Mockingbirds by svetsol/shutterstock.com
Starlings by Bob Comix
Finches by Stefanie Schubbert/shutterstock.com
Sparrows & Towhees by Noah Strycker/shutterstock.com
Snow Bunting by Gallinago_media/shutterstock.com
Tanagers, Cardinals, Grosbeaks, Buntings & Dickcissels based on a photo by Michael Fish
Blackbirds by Birchside/gograph.com
Wood-Warblers by thesilhouettequeen/123rf.com

INDEX

C

D

PHOTO CREDIT: CHRIS FLINK

DEXTER PATTERSON is also known as the Wisco Birder on social media. Passionate about making birdwatching inclusive for everyone, Dexter's lively "You Ready? Let's Go!" videos showcase his discovery of various bird species and have garnered millions of views. As an educator, photographer, writer, and cofounder of the BIPOC Birding Club of Wisconsin, Dexter is committed to demonstrating that birding is open to all. His goal is to promote inclusivity and help people connect with the beauty of nature. Through his efforts, he encourages a new wave of birdwatchers to venture into the outdoors and admire the diverse life surrounding them.